Erythrocyte - A Peripheral Biomarker For Infection and Inflammation

Edited by Kaneez Fatima Shad

Published in London, United Kingdom

IntechOpen

Supporting open minds since 2005

Erythrocyte - A Peripheral Biomarker For Infection and Inflammation
http://dx.doi.org/10.5772/intechopen.92510
Edited by Kaneez Fatima Shad

Contributors
Razan Zohairee, Marina Nagervadze, Marina Koridze, Rusudan Khukhunaishvili, Maria Sarkhaiani, Ketevan Dolidze, Leila Akhvlediani, Sophiko Tskvitinidze, Shorena Gabaidze, Irina Nakashidze, Sopio Garakanidze, Giorgi Nikolaishvili, Erland Johansson, Anders B. Falk, Kaneez Fatima Shad, Fahmida Abdi, Ann M. Simpson, Sara Lal, Vani Rajashekaraiah, Carl Hsieh, Masannagari Pallavi, Nazar Luqman Bilgrami

First published in London, United Kingdom, 2021 by IntechOpen
IntechOpen is the global imprint of INTECHOPEN LIMITED, registered in England and Wales, registration number: 11086078, 5 Princes Gate Court, London, SW7 2QJ, United Kingdom
Printed in Croatia

British Library Cataloguing-in-Publication Data
A catalogue record for this book is available from the British Library

Additional hard and PDF copies can be obtained from orders@intechopen.com

Erythrocyte - A Peripheral Biomarker For Infection and Inflammation
Edited by Kaneez Fatima Shad
p. cm.
Print ISBN 978-1-83969-120-1
Online ISBN 978-1-83969-121-8
eBook (PDF) ISBN 978-1-83969-122-5

For EU product safety concerns:
IN TECH d.o.o., Prolaz Marije Krucifikse Kozulić 3, 51000 Rijeka, Croatia, info@intechopen.com or visit our website at intechopen.com.

Meet the editor

Professor Kaneez Fatima Shad, an Australian Neuroscientist with a medical background, received a Ph.D. in 1994 from the Faculty of Medicine, UNSW, Australia, followed by a postdoc at the Allegheny University of Health Sciences, Philadelphia, USA. She taught Medical and Biological Sciences at various universities in Australia, the United States, United Arab Emirates, Bahrain, Pakistan, and Brunei. During this period, she was also engaged in research, obtaining local and international grants (for more than $2.3 million USD) to translate products such as a rapid diagnostic test for stroke and other vascular disorders (i.e., schizophrenia). Dr. Shad has published more than sixty articles in refereed journals, edited four books, written seven book chapters, presented at more than eighty-five international conferences, and mentored thirty-four postgraduate students. She is an international mentor and a protocol development specialist.

Contents

Preface XI

Section 1
Cell with a Signature 1

Chapter 1 3
Introductory Chapter: Anaemia and Iron Deficiency in Heart Failure
by Kaneez Fatima Shad and Nazar Luqman Bilgrami

Chapter 2 9
Erythrocytes as Biomarkers of Virus and Bacteria in View of Metal Ion Homeostasis
by Erland Johansson and Anders B. Falk

Chapter 3 33
Modulations in Oxidative Stress of Erythrocytes during Bacterial and Viral Infections
by Vani Rajashekaraiah, Carl Hsieh and Masannagari Pallavi

Section 2
Functional Biomarkers 53

Chapter 4 55
Early Predictive Biomarkers for Hypertension Using Human Fetal Astrocytes
by Fahmida Abdi, Ann M. Simpson, Sara Lal and Kaneez Fatima Shad

Chapter 5 77
Detection of Benzo[a]Pyrene Diol Epoxide-DNA Adducts in White Blood Cells of Asphalt Plant Workers in Syria
by Razan Zohairee

Chapter 6 87
The Study of Some Possible Risk Factors for Arterial Thrombosis in the Example of Georgian Patients
by Marina Koridze, Marina Nagervadze, Maria Sarkhaiani, Leila Akhvlediani, Rusudan Khukhunaishvili, Ketevan Dolidze, Sophiko Tskvitinidze, Shorena Gabaidze, Irina Nakashidze, Sopio Garakanidze and Giorgi Nikolaishvili

Preface

Erythrocytes or red blood cells (RBCs) make up about 84% of total cells in the human body and approximately 2.4 million RBCs are produced per second in an adult human. Each erythrocyte contains 270 million haemoglobin molecules responsible for carrying oxygen and giving red colour to these cells.

There are no nuclei, mitochondria, or any other organelles in mature RBCs. As such, they do not contain DNA and are not able to produce any RNA or protein, and thus cannot be attacked by viruses.

Although RBCs are considered bags for carrying haemoglobin for gas exchange, their cell membrane contains a number of lipids, proteins, and carbohydrates. This makes them capable of acting as peripheral biomarkers for many pathological conditions.

Early identification of key changes in erythrocytes in response to inflammatory or infectious diseases saves millions of lives worldwide.

Erythrocyte - A Peripheral Biomarker for Infection and Inflammation is divided into two sections: "Cells with a Signature" and "Functional Biomarkers."

The first chapter in Section 1 describes the role of iron deficiency (ID) in erythrocytes in regulating anaemia and heart failure. In heart failure, the ID rather than anaemia serves as the predictor for clinical outcome. In fact, ID has been shown to be a stronger predictor of the patients outcome than anaemia. ID without anaemia carries greater risk and poorer outcome compared to an anaemic patient without ID.

The second chapter in Section 1 discusses the role of erythrocytes in immunology, their interaction with viruses, metal ions, trace elements, and reactive oxygen species, and their involvement in sepsis and genetic disorders. It also examines the modulation of oxidative stress (OS) in erythrocytes in bacterial and viral infections. This chapter demonstrates how variations in RBC's proteins, lipids, and antioxidant capacity can be used as an OS biomarker to evaluate the efficiency of the erythrocytes in response to oxidative insults due to viral or bacterial intrusion. These infections alter the antioxidant capacity of erythrocytes, causing damage to surrounding cells and tissues. Thus, OS biomarkers can be used to gain insights into the effects of bacterial and viral infections on the erythrocyte microenvironment which is described in the third chapter.

Section 2 begins with a chapter describing how to measure early predictive biomarkers for hypertension by using human foetal astrocytes (HFAs) as an experimental model. The authors explain and demonstrate why and how HFAs are a good model for the detection of biomolecules that can predict the future onset of high blood pressure. Long processes of HFAs are mainly supported by intermediate filaments (IF), and glial fibrillary acidic protein (GFAP) is classified as a type III intermediate filament protein that is abundantly present in astrocytes. Elevated GFAP levels are being considered as a marker of astroglial injury, indicating the conversion of non-reactive (A2) into reactive (A1) astrocytes.

In this study, the authors also demonstrate that astrocytes from spontaneous hypertensive rats (SHRs) and their normal counterparts (WKY) rats have a remarkably similar profile of GFAP intensity to that of reactive (A1) and non-reactive (A2) HFAs, indicating that the HFA model can be used to study hypertension. They observed that reactive (A1) HFAs contain more calcium-activated proteins such as calpain, calpastatin, cathepsin, and mitogen-activated protein kinase (MAPK) as compared to normal (A2) HFAs, suggesting the possible link of these proteins with the future onset of HTN. Hence these proteins could be considered as potential candidates for predictive biomarkers of HTN. Similarly, DNA damage by polycyclic aromatic hydrocarbons was detected in the white blood cells (WBCs) of asphalt plant workers and evaluated by measuring the Benzo[a]pyrene diol epoxide (BPDE)-DNA in their WBCs. In the fifth chapter, the authors clearly show the carcinogenic effects of B[a]P on the workers in the asphalt company and that these BPDE-DNA adducts can act as biological markers for exposure risk assessment. Finally, in the last chapter the readers will be interested to know that there is an exponential increase in the risk of arterial and venous thrombotic events with age, gender, smoking habits, and diet types. The authors demonstrate an association of arterial thrombosis with ABO blood group, age, gender, smoking habit, and Rh, Kell, and MN blood groups in a population in Georgia.

I hope readers will achieve a better understanding of the various aspects of erythrocytes and their interaction with other cells in the blood and other tissues.

I would like to thank Author Service Manager Mr. Josip Knapic for his hard work, constant support, and help.

Kaneez Fatima Shad
School of Life Sciences,
UTS,
Sydney, Australia

Section 1

Cell with a Signature

Chapter 1

Introductory Chapter: Anaemia and Iron Deficiency in Heart Failure

Kaneez Fatima Shad and Nazar Luqman Bilgrami

1. Introduction

The condition of an erythrocyte could be able to affect the internal and external conditions of all the vital and non-vital organs of a human body. These little biconcave disk-shaped red structures of approximately 8 μm in diameter and of about 2.5 μm in thickness can reflect the extent and type of parasitic, bacterial, and viral infections such as malaria, diarrhoea, and covid 19. Lack of enough healthy red blood cells to carry oxygen results in anaemia. Anaemia is a condition that occurs due to reduced haemoglobin (Hb) in the blood. Thus, the oxygen-carrying capacity of the blood is reduced. When the oxygen-carrying capacity is reduced, the heart must work harder and faster to deliver the required oxygen to the body. If unrecognised, this process could result in serious damage to various organs of the body including the cardiovascular system.

Commonly encountered symptoms of anaemia are generalised weakness, fatigue, pale appearance, shortness of breath, irritable mood, lack of concentration etc. However, if gone untreated, anaemia could result in, precipitate, or aggravate heart failure (HF). It could also result in arrhythmias & precipitate heart attacks (type 2 myocardial infarction). Anaemia is usually caused by inefficient production of red blood cells or the haemoglobin within these cells; by the loss of blood due to haemolysis within the circulation or bleeding issues; or due to other chronic medical conditions like kidney disease or cancer. However, the most common reason for anaemia remains iron deficiency (ID) which is critical to produce Hb. ID could exist with and without anaemia.

2. Anaemia and iron deficiency in heart failure

Anaemia & ID are common in patients with heart failure. They are associated with poor clinical status and worse prognosis [1]. Compared to a 10% prevalence of anaemia in the general population it is found in 30% of patients with stable heart failure and 50% in those hospitalised, inclusive of heart failure with reduce ejection fraction (HFrEF) or heart failure with preserved ejection fraction (HFpEF) [2–4]. Although the pathogenesis of anaemia in heart failure is multifactorial including deficiency of vitamin B12 & folate, low erythropoietin production due to renal involvement in heart failure, inflammatory components suppressing the production of red blood cells and Hb, ID remains one of the major contributors [4–7].

In heart failure it is the ID rather than anaemia that serves as the predictor for clinical outcome [8]. impaired quality of life (QoL), reduced exercise capacity, greater morbidity, and mortality all could be predicted by the ID [8–10]. In fact ID has been shown to be a strong predictor for patient outcome than anaemia. ID without anaemia carries higher risk and poorer outcome compared to an anaemic patient without ID [8].

3. Management strategies

Earlier studies have suggested that anaemia is common in HF and carries worse prognosis yet measures to increase haemoglobin through red blood cell transfusion or erythropoiesis stimulating agents have not shown any beneficial effect in long term [4, 11, 12].

Unfortunately, this approach may not only did not improve the outcome in patients with HF but were associated with higher risk of adverse effects. This was demonstrated in a large study of public discharge database of 596456 patients admitted with heart failure [13]. Anaemia was present in 27% patient. The adjusted risk mortality was 70% higher in that receiving blood transfusion compared to 10% in those who did not. Thus, packed cell transfusion could be beneficial in acute anaemic states in heart failure but not in the chronic management.

As regards treatment with erythropoiesis stimulating agents (ESA), a meta-analysis of 11 earlier studies comprising 794 patients showed benefit in peak oxygen consumption, NYHA class, BNP levels and QoL but no significant effect on all-cause mortality [14]. Subsequently, a large study randomising 2278 patients with heart failure and anaemia but no iron deficiency when followed for 28 months did not confirm any benefit with the ESA darbopoetin, neither on the primary composite outcome of death from any cause, hospitalisation or worsening of heart failure. The lack of effect was equally present in subgroups as well as on the secondary outcomes of fatal MI, stroke, hypertension, and heart failure [15].

As against failure of blood replacement or ESA to demonstrate beneficial effect in heart failure, the scenario is different when it comes to treating ID with iron replacement. As mentioned earlier, ID is quite common in HF, affecting nearly 50% of patients with or without anaemia [8, 16–18]. Iron deficiency in patients with HF could be absolute or functional. Total iron is reduced in absolute ID but in the functional ID the total body iron is normal or increased but inadequate to fulfil the need of body tissues due to sequestration of the storage pools. Studies have shown that reversing ID in patient with or without anaemia and HF could improve patient outcomes [19], improve QoL and exercise capacity as well as improve depression which is also prevalent in patients with HF.

For the treatment of ID, it is important to consider the route of supplementation, whether oral of intravenous (IV). The dietary iron is absorbed in the duodenum by the enterocytes and subsequently taken up into circulation, bound to transferrin [7]. However, in HF there is gastrointestinal oedema leading to impairment of absorption of iron and duodenal iron transport [10]. Also, in HF the hepcidin level increases which further reduced the iron levels [20]. Several studies and meta-analysis data have studied the role if intravenous iron therapy in patients with HF and ID [21], showing its beneficial effect on NYHA class, QoL and 6 min walk test. FAIR-HF (Ferinject Assessment in Patients with Iron Deficiency and Chronic Heart Failure) randomised 459 patients with ID to IV iron or saline. IV iron therapy improved the QoL, 6 min walk test, NYHA level significantly but all-cause mortality or first hospitalisation remained unchanged [21]. However, a subsequent study [22] with similar recruitment criteria conducted on 304 patients but monitored for longer duration (52 weeks) showed significant improvement in primary end point of 6 min walk test, the benefit sustaining for 1 year. There was improvement in secondary end points NYHA class, QoL, fatigue score and there was reduction in the risk of hospitalisation for worsening of HF. However, the all-cause mortality did not change. Subsequent studies also showed lower rates of hospitalisation and cardiovascular mortality [21].

While advocating IV iron replacement in patients with HF one need be cautious with the potential adverse effects of iron overload. As iron is not excreted from the body in any form, the mechanism to control iron levels in the body is through regulation of its absorption through duodenum [23] which is bypassed when IV iron is given. Excess iron can cause cardiomyopathy, widespread tissue damage and endothelial dysfunction, all related with adverse cardiovascular outcomes [24]. It is therefore critical to prevent iron overload when correcting ID in patients with HF.

4. Conclusion

In this book, we are describing the novel and traditional approaches taken by different scientists around the world to relate the status of an erythrocyte (Red blood cell) with chronic and acute conditions such as heart failure, which can be happening slowly over a period or can happen suddenly.

Heart failure is associated with anaemia about 50% of the patients. Iron deficiency is one of the important caused of anaemia in HF. This is predominantly due to lack of absorption of iron from the duodenum that is often inflamed/congested in heart failure. Thus, iron replacement is pivotal in the management of HF with ID. Intravenous iron Administration of intravenous iron has been proved to be beneficial in improving symptoms as well as prognosis in patients with HF and ID.

Author details

Kaneez Fatima Shad[1*] and Nazar Luqman Bilgrami[2]

1 School of Life Sciences, University of Technology, Sydney, Australia

2 Austin Health, Heidelberg, Victoria, Australia

*Address all correspondence to: ftmshad@gmail.com

References

[1] Yancy CW, Jessup M, Bozkurt B, Butler J, Casey DE Jr, Colvin MM, Drazner MH, Filippatos GS, Fonarow GC, Givertz MM, Hollenberg SM, Lindenfeld J, Masoudi FA, McBride PE, Peterson PN, Stevenson LW, Westlake C. 2017 ACC/AHA/HFSA focused update of the 2013 ACCF/AHA guideline for the management of heart failure: a report of the American College of Cardiol- ogy/ American Heart Association Task Force on Clinical Practice Guidelines and the Heart Failure Society of America. Circulation. 2017;136: e137– e161.

[2] Anand IS. Anaemia and chronic heart failure implications and treatment options. J Am Coll Cardiol. 2008; 52:501-511.

[3] Tang YD, Katz SD. The prevalence of anaemia in chronic heart failure and its impact on the clinical outcomes. Heart Fail Rev. 2008; 13:387-392.

[4] Goodnough LT, Schrier SL. Evaluation and management of anaemia in the elderly. Am J Hematol. 2014; 89:88-96.

[5] Opasich C, Cazzola M, Scelsi L, De Feo S, Bosimini E, Lagioia R, Febo O, Ferrari R, Fucili A, Moratti R, Tramarin R, Tavazzi L. Blunted erythropoietin production and defective iron supply for erythropoiesis as major causes of anaemia in patients with chronic heart failure. Eur Heart J. 2005;26:2232– 2237.

[6] van der Meer P, Voors AA, Lipsic E, Smilde TD, van Gilst WH, van Veld- huisen DJ. Prognostic value of plasma erythropoietin on mortality in patients with chronic heart failure. J Am Coll Cardiol. 2004;44:63-67.

[7] Deswal A, Petersen NJ, Feldman AM, Young JB, White BG, Mann DL. Cytokines and cytokine receptors in advanced heart failure: an analysis of the cytokine database from the Vesnarinone trial (VEST). Circulation. 2001;103:2055-2059.

[8] Klip IT, Comin-Colet J, Voors AA, Ponikowski P, Enjuanes C, Banasiak W, et al. Iron deficiency in chronic heart failure: an international pooled analysis. Am Heart J. 2013;165(4):575-582 e3.

[9] Felker GM, Adams KF, Jr., Gattis WA, O'Connor CM. Anaemia as a risk factor and therapeutic target in heart failure. J Am Coll Cardiol. 2004;44(5):959-966.

[10] Anand IS, Gupta P. Anaemia and Iron Deficiency in Heart Failure: Current Concepts and Emerging Therapies. Circulation. 2018;138(1): 80-98.

[11] Mazer CD, Whitlock RP, Fergusson DA, Hall J, Belley-Cote E, Connolly K, Khanykin B, Gregory AJ, de Médicis É, McGuinness S, Royse A, Car- rier FM, Young PJ, Villar JC, Grocott HP, Seeberger MD, Fremes S, Lel- louche F, Syed S, Byrne K, Bagshaw SM, Hwang NC, Mehta C, Painter TW, Royse C, Verma S, Hare GMT, Cohen A, Thorpe KE, Jüni P, Shehata N; TRICS Investigators and Perioperative Anesthesia Clinical Trials Group. Restrictive or liberal red-cell transfusion for cardiac surgery. *N Engl J Med*. 2017;377:2133-2144.

[12] Kidney Disease Improving Global Outcomes) (KDIGO) Anaemia Work Group. KDIGO clinical practice guideline for anaemia in chronic kidney dis- ease. *Kidney Int Suppl*. 2012;2:279-335.

[13] Kao DP, Kreso E, Fonarow GC, Krantz MJ. Characteristics and out- comes among heart failure patients with anaemia and renal insuf- ficiency with and without blood transfusions (public discharge data from California 2000-2006). Am J Cardiol. 2011;107:69-73.

[14] Kotecha D, Ngo K, Walters JA, Manzano L, Palazzuoli A, Flather MD. Erythropoietin as a treatment of anaemia in heart failure: systematic re- view of randomized trials. Am Heart J. 2011;161:822-831.

[15] Swedberg K, Young JB, Anand IS, Cheng S, Desai AS, Diaz R, Maggioni AP, McMurray JJ, O'Connor C, Pfeffer MA, Solomon SD, Sun Y, Tendera M, van Veldhuisen DJ; RED-HF Committees; RED-HF Investigators. Treat- ment of anaemia with darbepoetin alfa in systolic heart failure. N Engl J Med. 2013;368: 1210-1219.

[16] Yeo TJ, Yeo PS, Ching-Chiew Wong R, Ong HY, Leong KT, Jaufeerally F, Sim D, Santhana krishnan R, Lim SL, M Y Chan M, Chai P, Low AF, Ling LH, Ng TP, Richards AM, Lam CS. Iron deficiency in a multi-ethnic Asian popu- lation with and without heart failure: prevalence, clinical correlates, func- tional significance and prognosis. Eur J Heart Fail. 2014;16:1125-1132.

[17] Martens P, Nijst P, Verbrugge FH, Smeets K, Dupont M, Mullens W. Im- pact of iron deficiency on exercise capacity and outcome in heart failure with reduced, mid-range and preserved ejection fraction. Acta Cardiol. 2017;73:1-9.

[18] Cappellini MD, Comin-Colet J, de Francisco A, Dignass A, Doehner W, Lam CS, Macdougall IC, Rogler G, Camaschella C, Kadir R, Kassebaum NJ, Spahn DR, Taher AT, Musallam KM; IRON CORE Group. Iron deficiency across chronic inflammatory conditions: international expert opinion on definition, diagnosis, and management. *Am J Hematol*. 2017;92:1068– 1078.

[19] Bolger AP, Bartlett FR, Penston HS, O'Leary J, Pollock N, Kaprielian R, et al. Intravenous iron alone for the treatment of anaemia in patients with chronic heart failure. J Am Coll Cardiol. 2006;48(6):1225-1227.

[20] Jankowska EA, Von Haehling S, Anker SD, MacDougall IC, Ponikowski P. Iron deficiency and heart failure: Diagnostic dilemmas and therapeutic perspectives. European Heart Journal. 2013;34(11):816-26c.

[21] Anker SD, Kirwan BA, van Veldhuisen DJ, Filippatos G, Comin-Colet J, Rus- chitzka F, Lüscher TF, Arutyunov GP, Motro M, Mori C, Roubert B, Pocock SJ, Ponikowski P. Effects of ferric carboxymaltose on hospitalisations and mor- tality rates in iron-deficient heart failure patients: an individual patient data meta-analysis. Eur J Heart Fail. 2018;20:125-133. doi: 10.1002/ejhf.823.

[22] Ponikowski P, van Veldhuisen DJ, Comin-Colet J, Ertl G, Komajda M, Mareev V, McDonagh T, Parkhomenko A, Tavazzi L, Levesque V, Mori C, Roubert B, Filippatos G, Ruschitzka F, Anker SD; CONFIRM-HF Investigators. Beneficial effects of long-term intravenous iron therapy with ferric carboxymaltose in patients with symptomatic heart fail- ure and iron deficiency. Eur Heart J. 2015;36:657-668. doi: 10.1093/eurheartj/ehu385.

[23] Ganz T, Nemeth E. Iron imports, IV: hepcidin and regulation of body iron metabolism. *Am J Physiol Gastrointest Liver Physiol*. 2006;290:G199– G203. doi: 10.1152/ajpgi.00412.2005.

[24] Sullivan JL. Long-term risks of increased use of intravenous iron. Lancet. 2007;370:481-482. doi: 10.1016/S0140-6736(07)61225-2.

Chapter 2

Erythrocytes as Biomarkers of Virus and Bacteria in View of Metal Ion Homeostasis

Erland Johansson and Anders B. Falk

Abstract

The erythrocyte contributes to the immune system in several ways. It sequesters interferons, interleukins or chemokines and by binding nucleic acid. It binds virus and bacteria and may deliver bacteria to macrophages for phagocytosis. It may also kill bacteria directly with oxygen. For proper function of the erythrocyte, homeostasis of reactive oxygen species, selenium, metal ions and trace elements is important. Erythrocytes display morphological and metabolic changes in diseases like sepsis, and in several genetic diseases. Patients infected with severe acute respiratory syndrome coronavirus 2 (SARS-CoV-2), giving rise to the coronavirus disease 2019 (Covid-19), show many erythrocyte changes as compared to healthy controls. The erythrocyte responds to hemolysins by purinergic signaling leading to hemolysis or phosphatidylserine exposure on the plasma membrane. Phosphatidylserine marks erythrocytes for clearance by spleen macrophages. Regulated erythrocyte cell death, also called eryptosis, can be induced by oxidative stress, pathogen infection, and certain diseases like sepsis. Erythrocytes may, in the future, contribute more to diagnosis based on research and diagnostic technological development.

Keywords: Erythrocyte, cytokine, sepsis, reactive oxygen species, selenium, iron, calcium, metal ions, ICP-MS, hierachy, Covid-19

1. Introduction

Recent research has shown that the erythrocyte contributes in many ways to defense against pathogens. The human erythrocyte is infected by the unicellular eukaryotic parasites *Plasmodium* and *Babesia* and the bacterium *Bartonella* [1]. Bacteria may adhere to glycosylated erythrocyte membrane proteins thereby facilitating phagocytosis by macrophages. Bacteria adhering to the erythrocyte surface may also be killed by oxygen release [2]. The erythrocyte also reacts with morphological changes to infections like Covid-19 or a disease like sepsis. However, the mammalian erythrocyte, having lost its nucleus and organelles, seems to play less a role in the immune system than erythrocytes of fish and birds [3]. The chapter will review aspects of mainly human erythrocytes in immunology and homeostasis with some special attention given to the ongoing Covid-19 pandemic.

2. The erythrocyte in immunology

Erythrocyte surface antigens play an important role in the erythrocyte interaction with pathogens and in other immune reactions. To begin with, the Duffy antigen has been found to be the entry point for *Plasmodium vivax*. This is supported by the finding that Duffy-negative individuals have some protection against *P. vivax* infection. Duffy negativity is common in the African population, although *P. vivax* may be able to infect some Duffy negative individuals [4]. Duffy was found to be a receptor of the chemokine with Cysteine-X-Cysteine (C-X-C) motif ligand 8 (CXCL8, also called interleukin-8) (**Figure 1**), and other chemokines such as chemokine with Cysteine-Cysteine (C-C) motif ligand 2 (CCL2, also called macrophage chemoattractant protein-1 (MCP-1)) were later also found to bind to Duffy [5]. The Duffy protein is a putative G-protein coupled receptor, but lacks some parts that would be necessary for signaling and is therefore regarded as a non-signaling receptor. The current hypothesis is that Duffy functions as a buffer-storage-sink for chemokines in the body. One indication for this is the reduced neutrophil counts and increased risk of acquiring human immunodeficiency virus (HIV-1) that has been found in Duffy-negative individuals [6]. However, overall neutrophil effector functions were largely unaffected in HIV-1-infected and non-infected Duffy-negative individuals [7]. One explanation for this could be an increased proteolytic activity as a compensatory mechanism for lower absolute neutrophil counts [7].

Chemokines belong to a group of immune signaling proteins called cytokines. In addition to chemokines, interferons, interleukins and the hormone erythropoietin are classified as cytokines. From the initial discovery of chemokine binding to Duffy, many more cytokines have been shown to be associated with the erythrocyte [8]. A considerable step forward was taken in a study where almost 50 different cytokines were identified in erythrocyte preparations [9]. Among chemokines, several from the C-C and the C-X-C families were found. Among interleukins, the interleukin-1, interleukin-6, interleukin-12 and interleukin-17 families were present (**Figure 1**). Present were also granulocyte-macrophage colony stimulating factor (GMCSF), interleukin-3, interleukin-5, tumor necrosis factor (TNF) superfamily and platelet derived growth factor (PDGF) (**Figure 1**). Interferon type I was represented by interferon-alpha2 whereas interferon type II was represented by interferon-gamma (**Figure 1**). The study suggests that the erythrocyte has significant capacity to contribute to the cytokine pool and immune homeostasis.

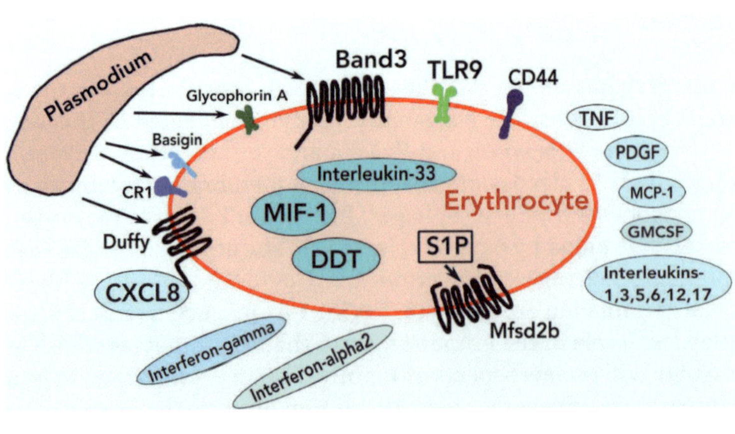

Figure 1.
Schematic figure of the erythrocyte in immunology. An erythrocyte contains 1.2 million band3 and 0.5 million glycophorin a molecules. Abbreviations are explained in the text.

In addition, cytokine and cytokine receptor gene expression has been found in both human and murine erythroblasts [10, 11].

Most of the cytokines have been localized extracellularly on the erythrocyte possibly bound to transmembrane receptors or associated to the glycocalyx. Macrophage migration inhibitory factor 1 (MIF1), D-dopachrome tautomerase (DDT) and interleukin-33 were found to have an intracellular localization [8, 12] (**Figure 1**). Interleukin-33 is a nuclear localized interleukin belonging to the interleukin-1 family [12]. Interleukin-33 is probably left over after expelling of the nucleus from the reticulocyte. MIF1 and DDT were found to exist in the erythrocyte, although the exact receptor has not been identified. In other tissues, a receptor for MIF1 and DDT is formed by CD74 and the hyaluronan receptor CD44 (**Figure 1**), which is also present on erythrocytes [13]. Intracellular as well as extracellular erythrocyte cytokines may be released by hemolysis or in protein-containing micro-vesicles or micro-particles produced by the erythrocyte [14–16].

Toll-like receptors (TLR) are a group of trans-membrane receptors of the innate immune system. TLR9 was shown to be present on erythrocytes [17] (**Figure 1**), although not all erythrocytes express TLR9 and considerable interindividual differences in expression was found [17]. Presence of mitochondrial but not nuclear DNA was confirmed on human and mouse erythrocytes. A knock-out mouse for TLR9 lost much of the mitochondrial DNA binding capacity otherwise found on human and mouse erythrocytes. TLR9 protein was mainly localized in the erythrocyte close to the plasma membrane probably binding to band3, a 14-transmembrane helices protein also known as anion exchanger 1 (AE1 or SLC4A1). Some evidence was found for erythrocyte scavenging of mitochondrial DNA from microvascular endothelial cells [17]. The erythrocyte may also interact with macrophages through the TLR9 receptor [18] (**Figure 1**). Additional TLRs may also exist in the erythrocyte, since expression of several TLRs was found in murine erythroblasts [11].

Erythrocytes also store and export the multifunctional molecule sphingosine-1-phosphate (S1P) (**Figure 1**). Inside the erythrocyte S1P has been shown to be involved in regulation of glycolysis [19]. S1P binds the S1P-receptor-1 on lymphocytes, antagonizing their egress from secondary lymphoid organs and the thymus. When bound to apolipoprotein M, S1P reduces lymphopoiesis by binding S1P-receptor-1 on common lymphoid progenitors in the bone marrow [20]. Erythrocytes synthesize S1P from sphingosine obtained from plasma in a reaction catalyzed by sphingosine-kinase-1. Because of lack of cellular organelles and their associated enzymes, no S1P-degrading enzyme is present in erythrocytes. These therefore effectively function as a S1P storage facility. Through site-directed mutagenesis and gene-knock-out studies, the major facilitator superfamily domain 2b (Mfsd2b) was found to be the exporter of S1P from erythrocytes [21]. Export of S1P from erythrocytes (**Figure 1**) was inhibited by an anion transport inhibitor, suggesting that also band3 may be involved in S1P export [22]. Erythrocytes are considered the major source of S1P plasma levels because of low plasma S1P levels in anemic patients. S1P plasma levels are also regulated by the lipid phosphate phosphatases, LPP1 and LPP3, on the plasma membrane of endothelial cells [23]. In addition, S1P levels in plasma may also depend on the presence and regulation of S1P carrier molecules like albumin and apolipoprotein M in complex with high-density lipoprotein.

Important for entry of *P. falciparum* into the erythrocyte are glycophorin A, band3, complement receptor 1 (CR1) and basigin [24] (**Figure 1**). Basigin is also known as the extracellular matrix metalloproteinase inducer (EMMPRIN) or CD147. The protein glycosylation underlying the ABO blood group system, is present on erythrocyte membrane proteins like band3. The ABO glycosylation is used by *Plasmodium* to facilitate rosetting, a stage in *Plasmodium* pathogenesis [25].

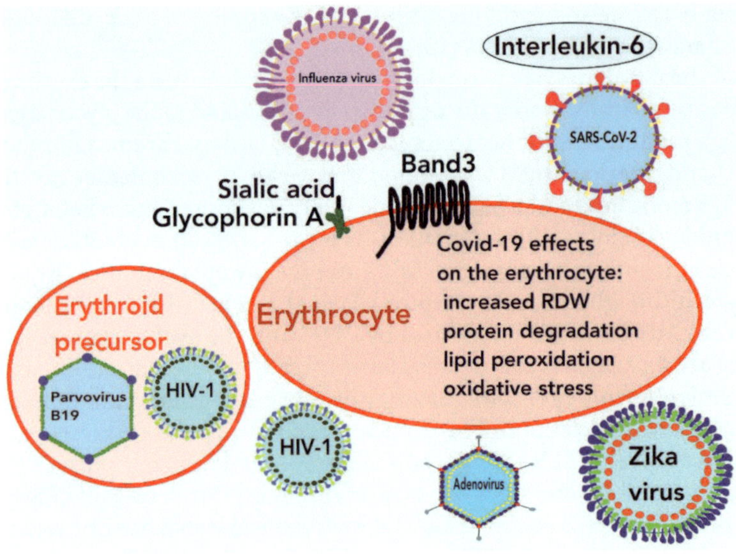

Figure 2.
Schematic figure of the erythrocyte and its interactions with viruses. Adenovirus and parvovirus lack a lipid bilayer around the virus capsid. Covid-19 disease affects the erythrocyte without entry of SARS-CoV-2 into the erythrocyte. Abbreviations are explained in the text.

Blood group O favors rosetting less than the A or B blood groups. Interestingly, the same pattern can be seen in Covid-19, where blood groups A or B are risk factors, whereas group O appears to be comparatively protective [26]. Glycophorin A is the most abundant sialo-glycoprotein on erythrocytes (**Figure 1**). Sialic acid residues are often used as receptors for invasion by bacteria and other pathogens [27] (**Figure 2**). Sequence analysis of glycophorin A from several primate species revealed high diversifying selection in the glycophorin A gene [27]. High selective pressure could be explained as a consequence of targeting of different pathogens by the glycophorin A receptor. This led to the pathogen decoy hypothesis, which states that the erythrocyte functions as a "flypaper", using glycophorin A to bind pathogens [27]. Pathogens may then be cleared by macrophages as erythrocytes pass through the spleen.

3. The erythrocyte and virus

The erythrocyte does not contain a nucleus or other organelles necessary for virus replication and survival. Its interaction with viruses is therefore mainly restricted to adherence of virus to the plasma membrane, transmembrane proteins or glycocalyx carbohydrates like sialic acid (**Figure 2**). Erythroid progenitor cells may however be infected with parvovirus B19 [28] which may lead to anemia. The ongoing Covid-19 pandemic has renewed interest in erythrocyte-virus interactions, due to the hypoxia or acute respiratory distress syndrome (ARDS) often seen in the disease [29]. Erythrocytes from Covid-19 patients showed increased protein degradation and glycolysis [29]. Increased or decreased levels of many lipids were seen [29]. Some signs of oxidative stress was noted, such as increased oxidized glutathione and increased protein levels of peroxiredoxin 2. Interestingly, N-terminal oxidation or degradation of band3 was noted (**Figure 2**). The cytosolic, N-terminal part of band3 has been shown to be of importance in signaling and hemoglobin regulation [30] and its oxidation or degradation may therefore explain some of the other noted effects of SARS-CoV-2 on the erythrocyte (**Figure 2**). Based on metabolic and molecular pathway analysis [29], protein degradation pathways, such as proteasome and ubiquitinylation components were identified as significantly affected in erythrocytes of Covid-19 patients (**Figure 2**). In addition, effects were noted on ferroptosis, cyclic-adenosine monophosphate (cAMP) and AMP-activated protein kinase signaling cascades. Lipid metabolism was also affected, especially acyl-carnitine and sphingolipid metabolism [29]. The effect on ferroptosis was based mainly on increased levels of peroxidated lipids and increased levels of heat shock protein beta-1, which inhibits ferroptosis by reducing intracellular levels of iron and maintaining levels of reduced glutathione [31]. The precise mechanism may be through inhibition of transferrin-receptor mediated iron uptake via stabilization of the cytoskeleton [31]. Ferroptosis is an iron-dependent non-apoptotic version of regulated cell death. It is caused by increased levels of iron accompanied by reduced activity of the selenium-dependent enzyme glutathione-peroxidase 4 and mitochondrial changes such as reduced mitochondrial size and mitochondrial membrane rupture. Protein levels of glutathione peroxidase 4 were unchanged in erythrocytes of Covid-19 patients [29]. The absence of mitochondria in the erythrocyte would argue against the existence of ferroptosis in the erythrocyte. In the Covid-19 infected patients, there were no alterations of directly clinically relevant hematological parameters, such as erythrocyte count, hematocrit, or mean corpuscular hemoglobin concentration [29]. The reason for hypoxia or ARDS in Covid-19 patients may therefore not be found in the erythrocyte, possibly with exception of the severe hypoxia without impaired lung function that has been seen in some patients [32]. Respiratory failure is the most common direct cause of death in Covid-19 patients [33, 34]. Studies have reported the nuclear remnants called Howell-Jolly bodies, or seemingly fully nucleated erythrocytes in Covid-19 patients [35, 36]. Nucleated erythrocytes indicate a bad prognosis and often occurs in critically ill patients in several diseases [37]. Mortality risk in Covid-19 seems to be associated with higher erythrocyte distribution width (RDW), a measure of heterogeneity of erythrocyte volume (**Figure 2**). In a meta-analysis of Covid-19 patients, higher RDW was associated with more severe disease [38]. The precise reason for this is unknown, but may involve reduced erythropoiesis or impaired erythrocyte clearance. A more prominent role of erythrocytes in the disease progression can therefore not be ruled out. For instance, in post-mortem analysis of Covid-19 patients with kidney injury, intact erythrocytes were found obstructing peritubular and glomerular capillary lumens [39]. A study of blood samples from Covid-19 patients showed depletion of several immune factors, among them interferon-alpha [40], an interferon that was previously localized in erythrocytes [9]. However, erythrocyte cytokine levels have so far not been analyzed in Covid-19 patients. Anemia in Covid-19 patients may be related to the increased interleukin-6 levels (**Figure 2**) that have been reported in the disease [41]. Interleukin-6 downregulates the iron exporter ferroportin via hepcidin [42]. Lower ferroportin activity makes iron stored in macrophages less available for erythropoiesis and for existing erythrocytes [43, 44]. Increased levels of interleukin-6 have also been implicated as being responsible for the cytokine release syndrome seen in some Covid-19 patients, particularly those affected by ARDS or severe hypoxia [41, 45]. The interleukin-6 receptor blocker tocilizumab showed disappointing results in Covid-19 patients [46, 47] suggesting a more complex pathophysiology of cytokine release syndrome in Covid-19 patients. In HIV-1 patients, anemia often appears together with secondary infections or nutritional deficiencies [44, 48], or is caused by treatment with zidovudine, which is a known cause of bone-marrow suppression [48]. Reduced erythropoietin levels in HIV-1-related anemia may have been caused by upregulation of proinflammatory cytokines like interleukin-1-beta and TNF-alpha [49] or by autoantibodies to erythropoietin [50]. HIV-1 may infect hematopoietic progenitor cells, but this is not considered a critical cause of anemia in HIV-1 patients [51].

Using biophysical modeling, band3 was suggested to be the point of attachment for SARS-CoV-2 on the erythrocyte surface [32]. The oxidation or degradation noted in band3 [29] may therefore have been caused directly by the binding of the virus. It should be noted that no entry of SARS-CoV-2 into the erythrocyte has as yet been proven (**Figure 2**). The effect of the virus on the erythrocyte may hence be due only to binding to surface receptors. The angiotensin-converting enzyme 2 (ACE2) membrane protein that is considered necessary for entry does not occur in erythrocyte membranes according to the erythrocyte proteome database (http://rbcc.hegelab.org/) [52], making ACE2 an unlikely entry or attachment point for SARS-CoV-2 on the erythrocyte surface [32]. Basigin, one of the erythrocyte receptors for *P. falciparum*, is not considered a receptor for SARS-CoV-2 [53], although such claims have been advanced [54]. The differential susceptibility conferred by the ABO blood group system to Covid-19 may imply a participation of the ABO glycosylation in attachment of the virus to the erythrocyte or other cells [26]. Band3, the suggested erythrocyte attachment protein for SARS-CoV-2, contains ABO glycosylation on the N-642 asparagine residue in the extracellular loop between transmembrane helices TM7 and TM8 [30]. This does not by itself necessarily imply involvement of the erythrocyte in Covid-19 pathogenesis since the ABO blood group system is expressed also on other membrane proteins and in other cells in the human body [55]. In contrast, the ABO blood group system seems not to be of importance for infection by the HIV-1 virus [56].

The erythrocyte has been suggested to be a hiding place for viruses, although these would not be able to infect the erythrocyte. Adsorption to erythrocytes has been shown for HIV-1, influenza and adenovirus [57–59] (**Figure 2**). Zika virus has been shown to survive in erythrocyte samples [60]. Zika virus could be detected in erythrocytes for a significantly longer time than in serum. Considering the ongoing SARS-CoV-2 pandemic, it would be of interest for applied transfusion and blood bank science to know the persistence of the SARS-CoV-2 virus on erythrocytes.

4. Erythrocyte selenium, metal ion and trace element status in relation to virus exposure

Metal ions, trace elements and other compounds contribute in different hierarchies to complex reactions that maintain blood homeostasis (**Figure 3**) [61]. Erythrocyte redox homeostasis is maintained through ascorbic acid, superoxide dismutase, catalase, glutathione and glutathione peroxidase. Antioxidant activity of selenium-dependent glutathione peroxidase may protect band3 of erythrocytes. Selenium is therefore important even for the basic oxygen and carbon dioxide exchange performed by the erythrocyte. Selenium as organic or inorganic form is metabolized by erythrocytes, platelets and neutrophils in different ways [62]. In one study, selenium was supplied together with plain food as inorganic sodium selenite at 200 microgram/day for one year. Neutrophils accumulated selenium more than platelets followed by erythrocytes [62]. In a similar study where selenium was supplied as 1-selenomethionine at 50 microgram/day for one year, platelets accumulated most selenium followed by erythrocytes and neutrophils [63]. These observations indicate that different cells metabolize inorganic and organic forms of selenium differently. The absence of nucleus and organelles in erythrocytes may partly explain the difference. Selenium supplied as inorganic sodium selenite was associated with decreased iron concentration in platelets and neutrophils.

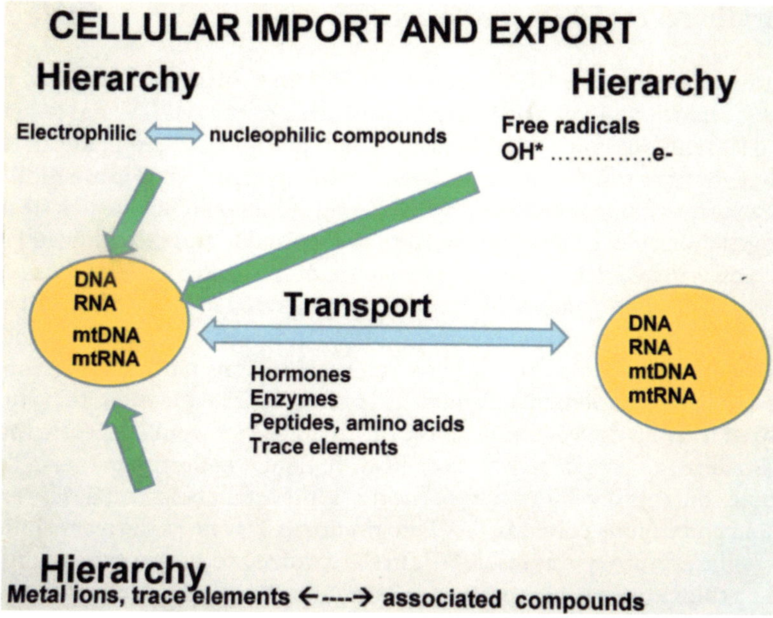

Figure 3.
The flow of compounds and metal ions between cells is dependent on the three main chemical pathways in a hierarchy. The first path involves free radical-induced production of compounds where hydroxyl radical and solvated electron reactions may be involved. The second path involves electrophilic and nucleophilic compounds forming products adequate to the cells. The third path involves metal ions and ligands dependent on the previous two pathways but adapted to cell demand. In erythroid precursors and in other cells, excluding the erythrocyte, compounds and metal ions may reach DNA and RNA not controlled by evolutionary developed genome adapted to cell demand. Small changes of DNA and RNA may take place, an epigenetic change, which may be restored or adapted, but large damages will present symptoms and are difficult to restore.

This result may indicate an iron-regulating function of selenium which may be particularly relevant for the erythrocyte considering its high iron content. Analysis of selenium and other elements was performed by micro-PIXE (micro-particle induced X-ray emission). Increased levels of Cd, Pb and Ag were reported in a study of erythrocytes in patients with Alzheimer's or Parkinson's disease, using the more sensitive and selective ICP-MS (inductively coupled plasma mass spectrometry) [61].

In view of these observations it is interesting that Covid-19 patients with higher selenium and selenoprotein P levels had lower mortality risk [64]. A correlation has also been found between cure rate and hair selenium concentration in Chinese Covid-19 patients [65]. As virus bind to erythrocytes, virus can be transported to different organs e.g. lung, kidney and brain. Viral particles may then find their way through the blood–brain-barrier or enter through the choroid plexus, an important part of the brain filtration system. The SARS-CoV-2 virus may decrease oxygen supply when attached to erythrocytes and impair brain oxygen supply. Low selenium intake may worsen the disease progress.

Erythropoiesis is partly regulated by microRNA and dependent on a high selenium status [66]. For instance, using a mouse loss-of-function allele, microRNA-142 was found to be required for maintaining typical erythrocyte shape, for normal metabolism of reactive oxygen species, and for overall erythrocyte lifespan [67]. MicroRNAs are differentially expressed depending on selenium status [68]. However, a precise role of selenium in regulating microRNA in relation to erythropoiesis has not been described. In addition, mechanisms underlying much of microRNA function are still not fully understood.

5. The erythrocyte and reactive oxygen species

Reactive oxygen species (ROS) are a set of oxygen containing compounds that easily react with and damage other constituents of the cell. Typical ROS in biological systems are superoxide, hydrogen peroxide and hydroxyl radical. ROS can be thought of as forming a hierarchy in the living organism (**Figure 3**). In the erythrocyte, superoxide is produced when Fe(II) of hemoglobin donates an electron to an oxygen molecule, generating the anionic superoxide and the oxidized Fe(III). This reaction is unavoidable due to the existence of iron and oxygen in close contact in the erythrocyte [69]. Superoxide can also be generated in the erythrocyte by nitric oxide synthase, an enzyme identical to the endothelial nitric oxide synthase (e-NOS). The superoxide is converted to water and hydrogen peroxide by copper and zinc-dependent superoxide dismutase. Hydrogen peroxide needs to be further metabolized to avoid the Fenton reaction which otherwise would generate the hydroxyl radical, a very reactive form of ROS. To defuse hydrogen peroxide the erythrocyte contains several enzymes, the most important being catalase, peroxiredoxin 2 and glutathione peroxidase 4. Peroxiredoxin 2 is one of the most abundant proteins in the erythrocyte cytosol [70]. In the oxidized form, peroxiredoxin 2 is recycled by thioredoxin using reduced nicotinamide adenine dinucleotide phosphate (NADPH) and thioredoxin reductase. Peroxiredoxin 2 is suggested to function as a non-catalytic scavenger of low-level hydrogen peroxide generated from autoxidation of hemoglobin [71]. However, a fraction of peroxiredoxin 2 is found attached to the erythrocyte plasma membrane and may be particularly important in detoxifying membrane lipid peroxides [72]. Catalase and the selenium-dependent glutathione peroxidase are important for the detoxification of hydrogen peroxide received by the erythrocyte from other parts of the vascular compartment. In plasma, superoxide is generated by xanthine oxidase, NADPH oxidase and nitric oxide synthase [73], enzymes that can occur in association to glycosaminoglycans of the endothelial cells of the vascular wall. Superoxide generated by these means can enter the erythrocyte by anion channels like band3. Alternatively, superoxide in plasma may be converted to hydrogen peroxide by extracellular superoxide dismutase. Hydrogen peroxide can then enter the erythrocyte by diffusion through the plasma membrane. Under stress conditions the erythrocyte receives much ROS generated by other cells inside or outside of the vascular compartment. Such ROS may have been generated by mitochondrial stress as a result of bacterial and viral infections (**Figure 3**). For instance, mitochondrial generation of ROS has been proposed as an important part of Covid-19 disease [74]. Antioxidant defense in erythrocytes also depends on small-molecule antioxidants like alpha-tocopherol, glutathione, ascorbate and reductants like NADPH generated in the pentose phosphate pathway. Alpha-tocopherol is present in erythrocyte plasma membrane where it may prevent lipid peroxidation. Glutathione participates as reductant in enzyme-catalyzed reactions. Ascorbate then recycles oxidized forms of alpha-tocopherol and glutathione back to the reduced forms. Superoxide can also react with nitric oxide generating peroxynitrite, a reactive nitrogen species [73]. The extent to which this reaction occurs depends on the presence of nitric oxide inside the erythrocyte. Nitric oxide is produced by nitric oxide synthase but also scavenged by oxyhemoglobin in the erythrocyte. Peroxynitrite can also be formed in plasma and may enter the erythrocyte as anion through band3. In conditions of stress, such as infection with SARS-CoV-2, ROS overload damage the erythrocyte plasma membrane which can lead to hemolysis. The necessity of the antioxidant system of erythrocytes has been proven in gene knock-out studies in mice. For instance, deletion of peroxiredoxin 1 or peroxiredoxin 2 led to increased plasma ROS, hemolytic anemia and shortened life and erythrocyte lifespan [75, 76].

6. The erythrocyte in sepsis

Sepsis is a strong and dysfunctional inflammatory response to an infection. In later stages it proceeds to immunosuppression, decreased pro-inflammatory cytokine levels and increased apoptosis of immune cells. The immunosuppressed stage can last for many months or even years and comes with increased risk for death due to secondary infections. Many aspects of sepsis have been highlighted recently like the presence of exosomes [77], complement and histones [78], metabolic changes [79] and ROS and inflammasomes [80]. Gene expression analysis of sepsis patients revealed CXCL8, tumor protein P53 and TNF as particularly important based on a subsequent pathway analysis [81, 82]. The chemokine CXCL8 binds to chemokine receptors on neutrophils, ultimately leading to attachment of neutrophils at sites of infection. Neutrophils have an important function in phagocytosis of bacteria through neutrophil extracellular traps (NET). Analysis of plasma protein levels of sepsis patients identified the cytokines interleukin-17 and interleukin-27 as being elevated early in sepsis, whereas interleukin-33 protein levels increased later in the immunosuppressive phase [83]. The three interleukins form an axis in sepsis pathophysiology with interleukin-27 having a inhibitory effect on the proinflammatory interleukin-17. The effect of interleukin-27 on interleukin-33 may depend on the stage of disease. Interleukin-33 has been shown to improve survival in murine sepsis models, which can be explained by increased neutrophil presence and more bacterial clearance at the site of infection. Interleukin-33 may have a protective effect by rebalancing different types of immunity [83]. Interleukin-33 has also been reported to suppress and modulate lymphocytes of the innate immune system such as type 2 innate lymphoid cells [84], and induce regulatory T cells. The contribution of erythrocytes to the pool of cytokines in sepsis has not been investigated. However, some of the cytokines implicated in sepsis pathophysiology like interleukin-17, interleukin-33 and CXCL8 have been reported to occur on erythrocytes from healthy individuals [9]. Pathophysiology of sepsis in individuals negative for the CXCL8 receptor Duffy or Duffy knock-out mice has not been reported. Apoptosis is a contributing factor to immunosuppression in sepsis [85]. A first hint of this was the finding that overexpression of B-cell lymphoma 2 (Bcl-2) protected against sepsis [86]. Bcl-2 is a mitochondrial protein that counteracts apoptotic processes. However, in the inflammatory phase of sepsis, some apoptosis of neutrophils is necessary [87].

Changed erythrocyte features have been reported in sepsis, such as the thorny erythrocytes called echinocytes [88] and increased RDW [89]. Reports have also claimed RDW to be of prognostic value in sepsis [90], but this has not been confirmed in other studies [91]. Erythrocytes seem to experience oxidative stress in sepsis [88, 89, 92]. Malondialdehyde, a product of lipid peroxidation, was increased by 3-fold in erythrocytes of sepsis patients [92]. Peroxynitrite has been reported as the major ROS in plasma of sepsis patients [88]. Sepsis patients often suffer from a moderate level of anemia [93]. Anemia in sepsis can have multiple causes such as anemia of chronic disease or hemolysis by bacterial toxins [43]. Hemolysis has often been reported in sepsis caused by *Clostridium perfringens* [94]. It is caused by disruption of the erythrocyte plasma membrane by phospholipase C activity of the *Clostridium* alpha-toxin [94]. A murine cecal ligation and puncture model of sepsis showed that sepsis is lethal in mice deficient for the heme-degrading enzyme heme oxygenase-1 [95]. The same study also showed that administration of the heme-sequestering protein hemopexin could prevent lethality. This suggests the importance of hemolysis in the pathogenesis of sepsis.

Sepsis can also be induced by virus [96]. For instance, COVID-19 and sepsis share similarities but also show some differences. Cytokine storm occurs in both diseases, although more so in the early stages of sepsis [97]. Lymphopenia has been reported in Covid-19 and is a typical feature of the immunosuppressed later stage of sepsis. Activation of the coagulation cascade occurs in both diseases, including reports of disseminated intravascular coagulation [98]. Similarities in pathophysiology implies that the same therapeutics may be relevant for both diseases. A murine model of sepsis based on cecal ligation and puncture recently showed increased survival with subcutaneous human interleukin-7 [99]. Positive results were later reported from a phase 2 clinical trial [100] of sepsis patients. Based on these results, interleukin-7 therapy has also been suggested for Covid-19 [97]. Interleukin-7 stimulates hematopoietic stem cells to differentiate into the common lymphoid progenitor. The rationale for interleukin-7 therapy is therefore a rebalancing of lymphoid cells, which is implicated by lymphopenia in both diseases. Interleukin-7 may also prevent apoptosis of immune cells [101]. Unfortunately, no approved sepsis-specific therapy exists despite many clinical trials.

7. Protein interactions and signaling mechanisms of possible immunological relevance

Band3 is the most abundant protein of the erythrocyte plasma membrane. The function of band3 is the excretion of bicarbonate formed from CO2 by carbonic anhydrase. A chloride ion is simultaneously imported. The process is reversed in the lungs where CO2 is released. Band3 has been implicated in intracellular signaling based on its cytosolic N-terminal domain, which consists of the first 360 of the 911 amino acid residues of the protein [30]. It was shown to interact with deoxyhemoglobin and enzymes of the glycolytic pathway. The much shorter C-terminal tail (amino acid residues 873–911) interacts with carbonic anhydrase II (CA) (**Figure 4**), forming a metabolon [102]. Band3 also interacts with ankyrin and spectrin [103], which stabilizes the band3 tetramer structure, calnexin during its maturation [30] and stomatin (**Figure 4**), which modulates the anion exchange activity of band3 [104]. Band3 often occurs in complex with glycophorin A presumably contributing to stabilization of the band3 tetramer in the plasma membrane [30]. In stored blood, band3 was found to interact also with flotillin-2, alpha-adducin and adenylosuccinate lyase [15].

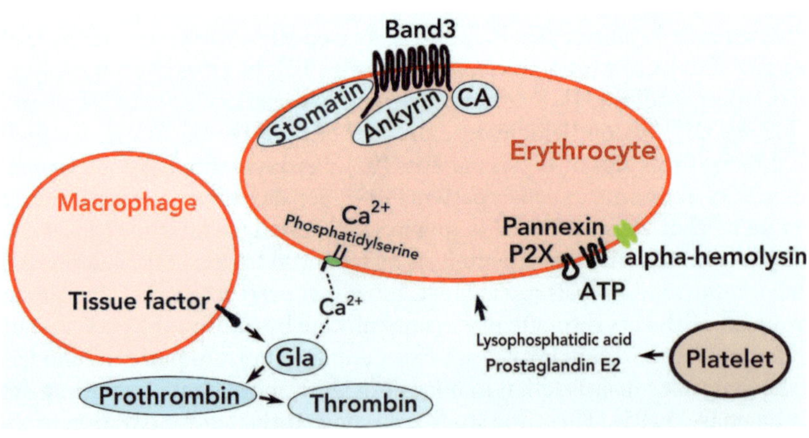

Figure 4.
Schematic figure of protein interactions and signaling mechanisms of the erythrocyte. Abbreviations are explained in the text.

A connection between blood coagulation and the immune system was shown by the finding that macrophages produce tissue factor, a transmembrane protein that starts the extrinsic coagulation cascade [105]. A connection between erythrocytes and coagulation is suggested by a correlation between decreased hematocrit and longer bleeding times. Erythrocytes receive signals in the form of lysophosphatidic acid and prostaglandin E2 from activated platelets [106] (**Figure 4**). Increasing cytosol calcium concentration then increases interaction forces between erythrocytes, thereby contributing to thrombus formation [107]. Increasing intracellular calcium levels also lead to phosphatidylserine exposure on the plasma membrane of the erythrocyte (**Figure 4**). Phosphatidylserine then acts as a binding surface for gamma-carboxyglutamyl residues of the vitamin K-dependent carboxylation/gamma-carboxyglutamic (Gla) domains of coagulation factors [108] (**Figure 4**). This initiates the formation of thrombin from prothrombin [109]. Contribution to coagulation also comes from erythrocyte micro-vesicles that similarly expose phosphatidylserine on the surface. However, erythrocyte micro-vesicles negatively regulate coagulation by supporting activated Protein C [110].

Erythrocytes respond to adenosine nucleotides by the adenosine tri-phosphate (ATP)-gated P2X and P2Y cation channel receptors in the erythrocyte plasma membrane. ATP is released from erythrocytes in an autocrine or paracrine mechanism of action in response to bacterial exotoxins called hemolysins [111]. They are produced by many bacteria and display several mechanisms of action. Alpha-hemolysin of *Escherichia coli* cause an increase in cellular calcium levels and opening of potassium and chloride ion channels [112]. ATP release then occurs either through pannexin channels (**Figure 4**) or through pores formed by hemolysins in the erythrocyte plasma membrane [111]. The subsequent activation of P2X receptors leads to exposure of phosphatidylserine on the surface of erythrocytes, making possible clearance by macrophages. Phosphatidylserine exposure can be prevented by P2X receptor blockers [112]. Purinergic signaling may be a way to avoid the damaging effects of intravascular hemolysis that would result without phosphatidylserine exposure and subsequent clearance by macrophages [112].

8. Senescence, aging, eryptosis and hemolysis of the erythrocyte

The erythrocyte has an average lifespan of around 120 days. As erythrocytes age they become less deformable, shrink in size and display dysfunctional hemoglobin accumulations. Calcium regulation may be part of what determines erythrocyte lifespan (**Figure 5**). Normally calcium concentrations are much lower in erythrocyte cytosol than in the surrounding fluid. During erythrocyte aging, intracellular levels of calcium increase. The events leading to this increase are not fully understood. One proposed mechanism is activation of the mechanosensitive PIEZO1 ion channel [113] through decreased deformability of the aging erythrocyte (**Figure 5**). PIEZO1 is a rather unspecific ion channel and can therefore cause general dissipation of ion gradients. In the case of calcium this would mean higher intracellular concentration. Gradually dysfunctional plasma membrane calcium efflux channels may also contribute to the higher calcium levels. Higher intracellular calcium concentrations are known to activate the Gardos channel, a potassium efflux channel in the plasma membrane involved in the regulation of cell volume in erythrocytes (**Figure 5**) and some other cells. Activation of the Gardos channel leads to potassium export and concomitant loss of water and chloride, together known as the Gardos effect. As they grow older, erythrocytes accordingly become more dehydrated. Support for this also comes from mutations in PIEZO1 [113] and the Gardos channel.

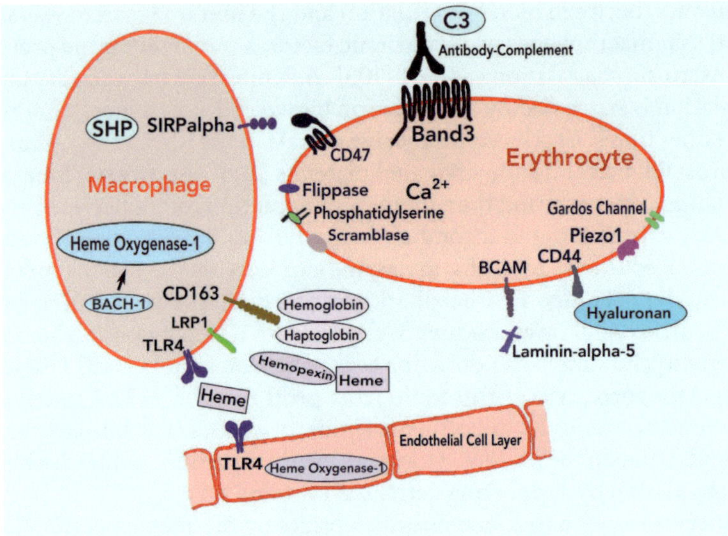

Figure 5.
Schematic figure of senescence, aging, eryptosis and hemolysis of the erythrocyte. Abbreviations are explained in the text.

The sequence of events from dehydration to erythrocyte clearance from circulation have not been fully elucidated. In one recently proposed model, erythrocytes may respond to dehydration by shedding Glycophorin C-containing vesicles, thereby losing membrane sialic acid [13]. Less sialic acid would result in activation of the basal cell adhesion molecule (BCAM, also known as Lutheran antigen) and CD44 membrane proteins of the erythrocyte. Ligands of BCAM and CD44 are laminin-alpha-5 and hyaluronan (**Figure 5**), known as parts of the extracellular matrix. Binding to laminin-alpha-5 or hyaluronan could possibly delay erythrocyte passage through spleen and liver making encounters with macrophages and clearance from circulation more likely.

Activation of the Gardos channel and dehydration are thus effects of increased erythrocyte calcium levels. Another consequence of increased calcium levels is phosphatidylserine exposure on the erythrocyte plasma membrane (**Figure 5**). This effect is likely to occur because of calcium-caused activation of scramblase, a membrane enzyme that equilibrates membrane lipids between the two plasma membrane leaflets. Further to this effect, calcium inactivates flippase, a membrane enzyme that moves lipids from the outer plasma membrane to the inner plasma membrane leaflet. Exposed phosphatidylserine can then bind to several phosphatidylserine receptors exposed on phagocytic cells [114], facilitating clearance of erythrocytes from circulation by spleen or liver macrophages. In young and healthy cells, phosphatidylserine occurs mainly on the inner leaflet of the plasma membrane. Exposure of phosphatidylserine is a marker for cell stress and effectively functions as an "eat-me" signal.

Yet another contribution to erythrocyte aging comes from the membrane protein CD47, a heavily glycosylated protein with five transmembrane domains and an extracellular N-terminal immunoglobulin-like domain (**Figure 5**). CD47 is a "don't eat me" signal that protects the cell from being phagocytosed by macrophages or other phagocytes. CD47 binds the macrophage signal-regulatory protein alpha (SIRP-alpha), a transmembrane protein with three immunoglobulin-like extracellular domains. CD47-SIRP-alpha interaction leads to phosphorylation of immunoreceptor tyrosine-based inhibition (ITIM) motifs in the cytoplasmic tail of SIRP-alpha. This leads to the recruitment of Src homology phosphatases (SHP) preventing myosin-IIA accumulation and phagocytosis by the macrophage. As erythrocytes grow older they display less CD47 on their surfaces, thereby increasing the likelihood for being eaten by macrophages. However, the role of CD47 in erythrocyte clearance may not be limited to a reduced "do not eat me" signal. CD47 can also bind thrombospondin-1, a multifunctional protein of the extracellular matrix. Evidence suggests that conformational changes in CD47, perhaps caused by oxidation as the cell ages, promotes binding of thrombospondin. The resulting complex then instead becomes an "eat me" signal when binding to SIRP-alpha of the macrophage [115].

Senescence and aging has been suggested to be regulated through band3 acting as a "molecular clock" for the erythrocyte [116]. Partial proteolysis, oxidation, phosphorylation and binding of methemoglobin or hemichrome to band3 leads to protein clustering and conformational changes that are recognized by anti-band3 autoantibodies (**Figure 5**). Binding of C3 complement then leads to phagocytosis of the aging erythrocyte when it encounters red pulp spleen or liver macrophages (**Figure 5**). Presumably, antibodies and complement must reach a threshold of opsonization for clearance by macrophages to take place. An appealing aspect of this "clock" is that it gauges the content of dysfunctional hemoglobin inside each erythrocyte, thus weeding out erythrocytes that have become unacceptably dysfunctional in their crucial role as oxygen carriers. If erythrocytes are ruled by a "clock", another candidate is the period circadian protein homolog 2 (PER2) of the circadian clock, a physiological system that determines the biological day-length of the organism. Erythrocytes of PER2-deficient mice showed morphological changes and impaired oxygen transport [117].

Eryptosis is the regulated cell death associated with erythrocytes. Eryptosis can be induced by oxidative stress, pathogen infection, certain diseases like sepsis and sickle-cell disease and certain drugs and chemicals. Eryptosis can also be protected against by several means, notably some plant-derived substances like ascorbate, caffeine and resveratrol [118]. Signaling leading to eryptosis involves ceramide, prostaglandin E2 and increased intracellular calcium levels. Similar to what happens in senescence and aging, this induces scramblase and inhibits flippase leading to phosphatidylserine exposure on the plasma membrane. Eryptosis is largely seen as a means for the organism to avoid the serious consequences of hemolysis and may even prevent pathogen growth [118]. Still, phosphatidylserine receptors on vascular endothelial cells can bind eryptotic erythrocytes leading to impaired microcirculation [119]. Regulation of eryptosis involves several kinases like the p21-activated kinase PAK2 and mitogen-and stress-activated kinase MSK1/2 [119]. Experiments with knock-out mice further suggested involvement of the beta-glucosidase-like protein Klotho. In healthy human and mice, Klotho is a co-receptor for fibroblast growth factor 23, regulating phosphate metabolism and vitamin-D synthesis. Klotho deletion ultimately leads to premature aging and death. Whether Klotho is involved in regulation of physiological eryptosis in humans remains unknown.

Both eryptosis and senescence involve erythrocyte-macrophage interactions. The close interactions between erythrocytes and macrophages follow throughout the erythrocyte lifespan. To begin with, macrophages participate in erythropoiesis, the formation and development of erythrocytes in the erythroblastic islands of the bone marrow. During the lifetime of the erythrocyte, passage through spleen and liver generates new macrophage encounters. In these encounters erythrocytes are screened for healthiness and may be repaired by the macrophages [120], for instance by removal of the inclusion bodies containing damaged hemoglobin, called Heinz bodies. Erythrocytes are the main source of blood MIF1 (macrophage migration inhibitory factor 1), although the physiological significance of this remains to be established [8]. At last, the senescent or eryptotic erythrocyte is phagocytosed by spleen or liver macrophages.

Hemolysis is the lysis of erythrocytes and release of their cellular contents in surrounding fluids. Intravascular hemolysis occurs mainly as a consequence of bacterial infection and production of endotoxins. Hemolysis may also occur as a consequence of shear stress when erythrocytes pass through narrow capillaries. The mechanism underlying hemolysis is overhydration due to defective ion efflux channels, particularly sodium/potassium ATPase. Cell volume increase will lead to rupture of the cell membrane and release of cytosolic contents. Hemolysis is typical of some diseases such as sickle-cell disease or glucose-6-phosphate deficiency. The fate of hemoglobin is an important aspect of hemolysis. Free hemoglobin typically binds to haptoglobin and the complex is then internalized through binding to the CD163 receptor on macrophages (**Figure 5**). If this does not occur, and if the iron has been oxidized to the ferric Fe(III) state, plasma hemoglobin will be separated into globin and heme. Being hydrophobic, free heme makes contact to plasma proteins such as albumin, alpha-1 antitrypsin, alpha-2 macroglobulin, lipoproteins and hemopexin. Binding to hemopexin effectively sequesters heme, making the complex largely safe for further circulation in the vascular compartment. The complex will eventually be internalized through binding to the low density lipoprotein receptor-related protein 1 (LRP1) receptor (**Figure 5**) on hepatocytes or macrophages [121]. If bound to albumin, heme can be transported into cells by binding to transferrin receptor 1. If cellular uptake does not happen by these routes, for instance due to hemopexin saturation, free heme can contribute to inflammation by binding the TLR4 receptor (**Figure 5**). Heme is in this context acting as a danger-associated molecular pattern (DAMP). Since TLR4 occurs also on endothelial cells, the vasculature is able to directly respond to increased heme levels. The effect of TLR4 stimulation is activation of NF-kB (nuclear factor kappa-light-chain-enhancer of activated B cells), inflammasome activation leading to activation of caspase-1 and release of pro-inflammatory interleukins like interleukin-1-beta, and interleukin-18. Recently, binding of heme to TLR4 was shown to be dependent on myeloid differentiation factor 2 (MD-2) and the glycosylphosphatidyl inositol (GPI)-anchored leucine-rich repeat protein CD14 [122]. As a comparison, in eryptosis and normal senescence and aging, macrophages typically phagocytose the whole erythrocyte and free heme does not appear.

Heme can also activate the alternative pathway of complement, and activate platelets through C-type-lectin-like receptor 2 [123]. In sickle cell disease patients and murine models of sickle-cell disease, hemolysis caused kidney deposits consisting of complement C3. The condition was probably caused by heme activation of complement, since it was possible to counteract by addition of hemopexin [124]. Free heme and its iron is available for use by bacterial pathogens and this is considered as an important aspect of hemolysis, for instance in hemolytic anemia. Sequestering of heme and hemoglobin is thus an important protection against damage caused by hemolysis. Inside the cell, heme is catabolized by heme oxygenase-1, producing biliverdin, free iron and carbon monoxide. The expression of heme oxygenase-1 can be induced by heme relieving gene repression through the BACH-1 heme-sensing repressor (**Figure 5**). The net effect of the catabolism is largely anti-inflammatory due to lower levels of TNF-alpha, interleukin-1-beta, and macrophage inflammatory protein (MIP) and upregulation of interleukin-10 [125]. The carbon monoxide produced as a by-product may however contribute to oxidative stress due to reactions with reduced transition metals [125]. Heme oxygenase-1 is itself a heme-containing enzyme of the heat shock protein family, present in the endoplasmic reticulum membrane oriented to the cytosol and expressed throughout the body (**Figure 5**). The globin part of hemoglobin is regarded as largely non-inflammatory [3]. Heme and its consequences are the most well-studied aspects of hemolysis. Less well-studied effects of hemolysis include release of intracellular or surface-bound cytokines from erythrocytes.

9. Some genetic diseases particularly relevant in the erythrocyte context

Glucose-6-phosphate dehydrogenase is an enzyme in the glycolysis and pentose phosphate pathways. Its deficiency particularly effects the erythrocyte since the erythrocyte depends on glycolysis and the pentose phosphate pathway for its energy needs. Erythrocyte irregularities called Heinz bodies are frequent in glucose-6-phosphate deficiency. The deficiency is particularly common in sub-Saharan Africa due to a survival advantage in malaria-afflicted areas. Interestingly, deficiency for glucose-6-phosphate dehydrogenase seems to be associated with hemolysis in Covid-19 patients [126]. Patients with glucose-6-phosphate deficiency may react with hemolysis as a response also to other infections and treatment with primaquine or hydroxychloroquine [127, 128], suggesting a more general susceptibility to hemolysis associated with glucose-6-phosphate deficiency. Some other diseases involving the erythrocyte are sickle-cell disease, thalassemia, stomatocytosis, spherocytosis, ovalocytosis, paroxysmal nocturnal hemoglobinuria (PNH), polycythemia vera, acute erythroid leukemia and lecithin-cholesterol acyltransferase deficiency (LCAT). Paroxysmal nocturnal hemoglobinuria (PNH) is an intravascular hemolytic anemia, most often caused by somatic mutations in the gene encoding phosphatidylinositol glycan A (PIGA), leading to dysfunctional GPI-dependent anchoring of membrane proteins. One such erythrocyte membrane protein is decay accelerating factor (DAF) which acts to limit action of the alternative complement pathway. The resulting complement-driven hemolysis can be counteracted by the humanized monoclonal antibody eculizumab directed against terminal complement component C5.

10. Outlook for new treatments and diagnosis

The Duffy antigen may be of diagnostic value for some diseases, particularly malaria, but may perhaps also prove to be a druggable target, since G-protein coupled receptors are among the most common drug targets. The signaling mechanisms discovered recently particularly regarding band3 may also be of pharmacological interest. The mechanism underlying eryptosis may be used to kill off infected erythrocytes [129]. It is known that antioxidants like ascorbic acid and catechin, provided through the diet can have protective effects on the erythrocyte. Refined erythrocyte related diet recommendations could perhaps be expected. The erythrocyte has previously been used in various physical or biochemical ways as a diagnostic tool. Such tests are the erythrocyte sedimentation rate, hematocrit, erythrocyte distribution width, erythrocyte count, mean corpuscular hemoglobin concentration and mean corpuscular volume. New diagnostic or treatment developments can be expected based on molecular, biotechnological and genetic techniques. One example would be ICP-MS to detect element profile changes of genetic, epigenetic or environmental origin.

Author details

Erland Johansson* and Anders B. Falk
EJSelenkonsult AB., Uppsala, Sweden

*Address all correspondence to: 101jejohansson@gmail.com

References

[1] McCullough J. RBCs as targets of infection. Hematology Am Soc Hematol Educ Program. 2014;**2014**(1):404-409.

[2] Minasyan H. Phagocytosis and oxycytosis: Two arms of human innate immunity. Immunol Res. 2018;**66**(2):271-280.

[3] Anderson HL, Brodsky IE, Mangalmurti NS. The evolving erythrocyte: Red blood cells as modulators of innate immunity. J Immunol. 2018;**201**(5):1343-1351.

[4] Popovici J, Roesch C, Rougeron V. The enigmatic mechanisms by which plasmodium vivax infects Duffy-negative individuals. PLoS Pathog. 2020;**16**(2):e1008258.

[5] Horuk R. The Duffy antigen receptor for chemokines DARC/ACKR1. Front Immunol. 2015;**6**:279.

[6] Reich D, Nalls MA, Kao WHL, Akylbekova EL, Tandon A, Patterson N, et al. Reduced neutrophil count in people of African descent is due to a regulatory variant in the Duffy antigen receptor for chemokines gene. PLoS Genet. 2009;**5**(1):e1000360.

[7] Naidoo KK, Ngubane A, Gaza P, Moodley A, Ndung'u T, Thobakgale CF. Neutrophil effector functions are not impaired in Duffy antigen receptor for chemokines (DARC)-null black south Africans. Front Immunol. 2019;**10**:551.

[8] Karsten E, Herbert BR. The emerging role of red blood cells in cytokine signalling and modulating immune cells. Blood Rev. 2020;**41**:100644.

[9] Karsten E, Breen E, Herbert BR. Red blood cells are dynamic reservoirs of cytokines. Sci Rep. 2018;**8**(1):3101.

[10] An X, Schulz VP, Li J, Wu K, Liu J, Xue F, et al. Global transcriptome analyses of human and murine terminal erythroid differentiation. Blood. 2014;**123**(22):3466-3477.

[11] Liang R, Campreciós G, Kou Y, McGrath K, Nowak R, Catherman S, et al. A systems approach identifies essential FOXO3 functions at key steps of terminal erythropoiesis. PLoS Genet. 2015;**11**(10):e1005526.

[12] Wei J, Zhao J, Schrott V, Zhang Y, Gladwin M, Bullock G, et al. Red blood cells store and release Interleukin-33. J Investig Med. 2015;**63**(6):806-810.

[13] Klei TRL, Dalimot JJ, Beuger BM, Veldthuis M, Ichou FA, Verkuijlen PJJH, et al. The Gardos effect drives erythrocyte senescence and leads to Lu/BCAM and CD44 adhesion molecule activation. Blood Advances. 2020;**4**(24):6218-6229.

[14] Li K-Y, Zheng L, Wang Q, Hu Y-W. Characteristics of erythrocyte-derived microvesicles and its relation with atherosclerosis. Atherosclerosis. 2016;**255**:140-144.

[15] Prudent M, Delobel J, Hübner A, Benay C, Lion N, Tissot J-D. Proteomics of stored red blood cell membrane and storage-induced microvesicles reveals the Association of Flotillin-2 with band 3 complexes. Front Physiol. 2018;**9**:421.

[16] Mantel P-Y. Malaria-infected erythrocyte-derived microvesicles mediate cellular communication within the parasite population and with the host immune system. Cell Host & Microbe. 2013;**13**:521-534.

[17] Hotz MJ, Qing D, Shashaty MGS, Zhang P, Faust H, Sondheimer N, et al. Red blood cells Homeostatically bind mitochondrial DNA through TLR9 to maintain quiescence and to prevent lung injury. Am J Respir Crit Care Med. 2018;**197**(4):470-480.

[18] Kang Y-Q, Yuan X-H, Li Z-Z, Wang H, Zhou X-F, Wang X-X, et al. Antishock characteristics of erythrocyte-mediated endoplasmic reticulum stress in macrophages in severe hemorrhagic shock environment based on TLR9-cGAS-STING-IFN signal Axis. Cell Transplant. 2020;**29**:0963689720950218.

[19] Sun K, Zhang Y, D'Alessandro A, Nemkov T, Song A, Wu H, et al. Sphingosine-1-phosphate promotes erythrocyte glycolysis and oxygen release for adaptation to high-altitude hypoxia. Nat Commun. 2016;**7**(1):12086.

[20] Blaho VA, Galvani S, Engelbrecht E, Liu C, Kono M, Proia RL, et al. HDL-bound sphingosine-1-phosphate restrains lymphopoiesis and neuroinflammation. Nature. 2015;**523**: 342-346.

[21] Vu TM, Ishizu A-N, Foo JC, Toh XR, Zhang F, Whee DM, et al. Mfsd2b is essential for the sphingosine-1-phosphate export in erythrocytes and platelets. Nature. 2017;**550**(7677): 524-528.

[22] Kurano M, Nishikawa M, Kuma H, Jona M, Yatomi Y. Involvement of Band3 in the efflux of sphingosine 1-phosphate from erythrocytes. PLoS One. 2017;**12**(5):e0177543.

[23] Dhangadamajhi G. Sphingosine 1-phosphate in malaria pathogenesis and its implication in therapeutic opportunities. Frontiers in Cellular and Infection Microbiology. 2020;**10**:14.

[24] Crosnier C, Bustamante LY, Bartholdson SJ, Bei AK, Theron M, Uchikawa M, et al. Basigin is a receptor essential for erythrocyte invasion by plasmodium falciparum. Nature. 2011;**480**(7378):534-537.

[25] McQuaid F, Rowe JA. Rosetting revisited: A critical look at the evidence for host erythrocyte receptors in plasmodium falciparum rosetting. Parasitology. 2020;**147**(1):1-11.

[26] Liu N, Zhang T, Ma L, Zhang H, Wang H, Wei W, et al. The impact of ABO blood group on COVID-19 infection risk and mortality: A systematic review and meta-analysis. Blood Rev. 2020;:100785.

[27] Baum J, Ward RH, Conway DJ. Natural selection on the erythrocyte surface. Molecular Biology and Evolution. 2002;**19**(3):223-229.

[28] Ganaie SS, Qiu J. Recent advances in replication and infection of human parvovirus B19. Front Cell Infect Microbiol. 2018;**8**:166.

[29] Thomas T, Stefanoni D, Dzieciatkowska M, Issaian A, Nemkov T, Hill RC, et al. Evidence of structural protein damage and membrane lipid remodeling in red blood cells from COVID-19 patients. J Proteome Res. 2020;**19**(11):4455-4469.

[30] Reithmeier RAF, Casey JR, Kalli AC, Sansom MSP, Alguel Y, Iwata S. Band 3, the human red cell chloride/bicarbonate anion exchanger (AE1, SLC4A1), in a structural context. Biochim Biophys Acta. 2016;**1858**:1507-1532.

[31] Lei P, Bai T, Sun Y. Mechanisms of Ferroptosis and relations with regulated cell death: A review. Front Physiol. 2019;**10**:139.

[32] Cosic I, Cosic D, Loncarevic I. RRM prediction of erythrocyte Band3 protein as alternative receptor for SARS-CoV-2 virus. Applied Sciences 2020;**10**(11):4053.

[33] Zhang B, Zhou X, Qiu Y, Song Y, Feng F, Feng J, et al. Clinical characteristics of 82 cases of death from COVID-19. PLoS ONE. 2020;**15**(7): e0235458.

[34] Berenguer J, Ryan P, Rodríguez-Baño J, Jarrín I, Carratalà J, Pachón J, et al. Characteristics and predictors of death among 4035 consecutively hospitalized patients with COVID-19 in Spain. Clinical Microbiology and Infection. 2020;**26**(11):1525-1536.

[35] Reva I, Yamamoto T, Rasskazova M, Lemeshko T, Usov V, Krasnikov Y, et al. Erythrocytes as a target of SARS CoV-2 in pathogenesis of Covid-19. AREM 2020;**10**(3):5-11.

[36] Jones JR, Ireland R. Morphological changes in a case of SARS-CoV-2 infection. Blood. 2020;**135**(25): 2324-2324.

[37] Purtle SW, Horkan CM, Moromizato T, Gibbons FK, Christopher KB. Nucleated red blood cells, critical illness survivors and postdischarge outcomes: A cohort study. Crit Care. 2017;**21**(1):154.

[38] Zinellu A, Mangoni AA. Red blood cell distribution width, disease severity, and mortality in hospitalized patients with SARS-CoV-2 infection: A systematic review and meta-analysis. J Clin Med. 2021;**10**(2):286.

[39] Su H, Yang M, Wan C, Yi L-X, Tang F, Zhu H-Y, et al. Renal histopathological analysis of 26 postmortem findings of patients with COVID-19 in China. Kidney International. 2020;**98**(1):219-227.

[40] Combes AJ, Courau T, Kuhn NF, Hu KH, Ray A, Chen WS, et al. Global absence and targeting of protective immune states in severe COVID-19. Nature. 2021. doi:10.1038/s41586-021-03234-7.

[41] Liu B, Li M, Zhou Z, Guan X, Xiang Y. Can we use interleukin-6 (IL-6) blockade for coronavirus disease 2019 (COVID-19)-induced cytokine release syndrome (CRS)? J Autoimmun. 2020;**111**:102452.

[42] Nemeth E, Rivera S, Gabayan V, Keller C, Taudorf S, Pedersen BK, et al. IL-6 mediates hypoferremia of inflammation by inducing the synthesis of the iron regulatory hormone hepcidin. J Clin Invest. 2004;**113**(9): 1271-1276.

[43] Weiss G, Goodnough LT. Anemia of chronic disease. N Engl J Med. 2005;**352**(10):1011-1023.

[44] Kerkhoff AD, Wood R, Cobelens FG, Gupta-Wright A, Bekker L-G, Lawn SD. Resolution of anaemia in a cohort of HIV-infected patients with a high prevalence and incidence of tuberculosis receiving antiretroviral therapy in South Africa. BMC Infectious Diseases. 2014;**14**(1):3860.

[45] Guillén L, Padilla S, Fernández M, Agulló V, García JA, Telenti G, et al. Preemptive interleukin-6 blockade in patients with COVID-19. Sci Rep. 2020;**10**(1):16826.

[46] Kow CS, Hasan SS. Interleukin-6 blockade with tocilizumab in COVID-19: Does it live up to its hype? Pulmonology. 2021;**27**(1):86-87.

[47] Cortegiani A, Ippolito M, Greco M, Granone V, Protti A, Gregoretti C, et al. Rationale and evidence on the use of tocilizumab in COVID-19: A systematic review. Pulmonology. 2021;**27**(1):52-66.

[48] Harding BN, Whitney BM, Nance RM, Ruderman SA, Crane HM, Burkholder G, et al. Anemia risk factors among people living with HIV across the United States in the current treatment era: A clinical cohort study. BMC Infect Dis. 2020;**20**(1):238.

[49] Camacho J, Poveda F, Zamorano AF, Valencia ME, Vázquez JJ, Arnalich F. Serum erythropoietin levels in anaemic

patients with advanced human immunodeficiency virus infection. Br J Haematol. 1992;**82**(3):608-614.

[50] Sipsas NV, Kokori SI, Ioannidis JP, Kyriaki D, Tzioufas AG, Kordossis T. Circulating autoantibodies to erythropoietin are associated with human immunodeficiency virus type 1-related anemia. J Infect Dis. 1999;**180**(6):2044-2047.

[51] Gibellini D, Clò A, Morini S, Miserocchi A, Ponti C, Re MC. Effects of human immunodeficiency virus on the erythrocyte and megakaryocyte lineages. World J Virol. 2013;**2**(2): 91-101.

[52] Hegedűs T, Chaubey PM, Várady G, Szabó E, Sarankó H, Hofstetter L, et al. Inconsistencies in the red blood cell membrane proteome analysis: generation of a database for research and diagnostic applications. Database (Oxford). 2015;**2015**:bav056.

[53] Shilts J, Wright GJ. No evidence for basigin/CD147 as a direct SARS-CoV-2 spike binding receptor. bioRxiv. 2020;2020.07.25.221036.

[54] Wang K, Chen W, Zhou Y-S, Lian J-Q, Zhang Z, Du P, et al. SARS-CoV-2 invades host cells via a novel route: CD147-spike protein. bioRxiv 2020;2020.03.14.988345.

[55] Sano R, Nakajima T, Takahashi Y, Kubo R, Kobayashi M, Takahashi K, et al. Epithelial expression of human ABO blood group genes is dependent upon a downstream regulatory element functioning through an epithelial cell-specific transcription factor, Elf5 *. Journal of Biological Chemistry. 2016;**291**(43):22594-22606.

[56] Motswaledi MS, Kasvosve I, Oguntibeju OO. The role of red blood cells in enhancing or preventing HIV infection and other diseases. Biomed Res Int. 2013;**2013**:758682.

[57] Beck Z, Brown BK, Wieczorek L, Peachman KK, Matyas GR, Polonis VR, et al. Human erythrocytes selectively bind and enrich infectious HIV-1 virions. PLoS One. 2009;**4**(12):e8297.

[58] Makkoch J, Prachayangprecha S, Payungporn S, Chieochansin T, Songserm T, Amonsin A, et al. Erythrocyte binding preference of human pandemic influenza virus a and its effect on antibody response detection. Ann Lab Med. 2012;**32**(4):276-282.

[59] Carlisle RC, Di Y, Cerny AM, Sonnen AF-P, Sim RB, Green NK, et al. Human erythrocytes bind and inactivate type 5 adenovirus by presenting Coxsackie virus-adenovirus receptor and complement receptor 1. Blood. 2009;**113**(9):1909-1918.

[60] Stone M, Bakkour S, Lanteri MC, Brambilla D, Simmons G, Bruhn R, et al. Zika virus RNA and IgM persistence in blood compartments and body fluids: A prospective observational study. Lancet Infect Dis. 2020;**20**(12):1446-1456.

[61] Johansson E, Westermarck T, Ek P, Latvus A, Atroshi F. Comparison of Erythrocytes for Individual Indications of Metabolism Changes in Parkinson's and Alzheimer's Diseases. In: Atroshi F, editor. Personalized Medicine, in Relation to Redox State, Diet and Lifestyle. IntechOpen. 2020 doi:10.5772/intechopen.91660.

[62] Johansson, E, Lindh, U. Availability of Selenium Compounds in Man. Proc 3rd Int Symp Ind uses Selenium Tellurium. 1984: 528-536.

[63] Johansson, E, Lindh, U. Interactions of selenium with metal ions at the cellular level. Biological Trace Element Research. 1987;**12**:101-108.

[64] Moghaddam A, Heller RA, Sun Q, Seelig J, Cherkezov A, Seibert L, et al. Selenium deficiency is associated with

mortality risk from COVID-19. Nutrients. 2020;**12**(7):2098.

[65] Zhang J, Saad R, Taylor EW, Rayman MP. Selenium and selenoproteins in viral infection with potential relevance to COVID-19. Redox Biology. 2020;**37**:101715.

[66] Liao C, Carlson BA, Paulson RF, Prabhu KS. The intricate role of selenium and selenoproteins in erythropoiesis. Free Radic Biol Med. 2018;**127**:165-171.

[67] Rivkin N, Chapnik E, Mildner A, Barshtein G, Porat Z, Kartvelishvily E, et al. Erythrocyte survival is controlled by microRNA-142. Haematologica. 2017;**102**(4):676-685.

[68] Feng Y, Xing Y, Liu Z, Yang G, Niu X, Gao D. Integrated analysis of microRNA and mRNA expression profiles in rats with selenium deficiency and identification of associated miRNA-mRNA network. Sci Rep. 2018;**8**(1): 6601.

[69] Rifkind JM, Mohanty JG, Nagababu E. The pathophysiology of extracellular hemoglobin associated with enhanced oxidative reactions. Front Physiol. 2015;**5**:500.

[70] Kakhniashvili DG, Bulla LA, Goodman SR. The human erythrocyte proteome: Analysis by ion trap mass spectrometry. Mol Cell Proteomics. 2004;**3**(5):501-509.

[71] Low FM, Hampton MB, Peskin AV, Winterbourn CC. Peroxiredoxin 2 functions as a noncatalytic scavenger of low-level hydrogen peroxide in the erythrocyte. Blood. 2007;**109**(6): 2611-2617.

[72] Kuhn V, Diederich L, Keller TCS, Kramer CM, Lückstädt W, Panknin C, et al. Red blood cell function and dysfunction: Redox regulation, nitric oxide metabolism, Anemia. Antioxid Redox Signal. 2017;**26**(13):718-742.

[73] Romero N, Denicola A, Radi R. Red blood cells in the metabolism of nitric oxide-derived peroxynitrite. IUBMB Life. 2006;**58**(10):572-580.

[74] Pagano G, Manfredi C, Pallardó FV, Lyakhovich A, Tiano L, Trifuoggi M. Potential roles of mitochondrial cofactors in the adjuvant mitigation of proinflammatory acute infections, as in the case of sepsis and COVID-19 pneumonia. Inflamm Res. 2021;**70**(2): 159-170.

[75] Lee T-H. Peroxiredoxin II is essential for sustaining life span of erythrocytes in mice. Blood. 2003;**101**(12):5033-5038.

[76] Neumann CA, Krause DS, Carman CV, Das S, Dubey DP, Abraham JL, et al. Essential role for the peroxiredoxin Prdx1 in erythrocyte antioxidant defence and tumour suppression. Nature. 2003;**424**(6948): 561-565.

[77] Murao A, Brenner M, Aziz M, Wang P. Exosomes in sepsis. Front Immunol. 2020;**11**:2140.

[78] Zetoune FS, Ward PA. Role of complement and histones in sepsis. Front Med. 2020;**7**:616957.

[79] McBride MA, Owen AM, Stothers CL, Hernandez A, Luan L, Burelbach KR, et al. The metabolic basis of immune dysfunction following sepsis and trauma. Front Immunol. 2020;**11**:1043.

[80] Zhao S, Chen F, Yin Q, Wang D, Han W, Zhang Y. Reactive oxygen species interact with NLRP3 Inflammasomes and are involved in the inflammation of sepsis: From mechanism to treatment of progression. Front Physiol. 2020;**11**:571810.

[81] Sweeney TE, Perumal TM, Henao R, Nichols M, Howrylak JA, Choi AM, et al. A community approach to mortality prediction in sepsis via gene expression analysis. Nat Commun. 2018;**9**(1):694.

[82] Stanski NL, Wong HR. Prognostic and predictive enrichment in sepsis. Nat Rev Nephrol. 2020;**16**(1):20-31.

[83] Morrow KN, Coopersmith CM, Ford ML. IL-17, IL-27, and IL-33: A novel Axis linked to immunological dysfunction during sepsis. Front Immunol. 2019;**10**:1982.

[84] Chun TT, Chung C-S, Fallon EA, Hutchins NA, Clarke E, Rossi A-L, et al. Group 2 innate lymphoid cells (ILC2s) are key mediators of the inflammatory response in Polymicrobial sepsis. Am J Pathol. 2018;**188**(9): 2097-2108.

[85] Cao C, Yu M, Chai Y. Pathological alteration and therapeutic implications of sepsis-induced immune cell apoptosis. Cell Death Dis. 2019;**10**(10):782.

[86] Hotchkiss RS, Swanson PE, Knudson CM, Chang KC, Cobb JP, Osborne DF, et al. Overexpression of Bcl-2 in transgenic mice decreases apoptosis and improves survival in sepsis. The Journal of Immunology. 1999;**162**(7):4148-4156.

[87] Zhang CY, Dong X, Gao J, Lin W, Liu Z, Wang Z. Nanoparticle-induced neutrophil apoptosis increases survival in sepsis and alleviates neurological damage in stroke. Sci Adv. 2019;**5**(11):eaax7964.

[88] Oliveira YPA de, pontes-de-Carvalho LC, Couto RD, Noronha-Dutra AA. Oxidative stress in sepsis. Possible production of free radicals through an erythrocyte-mediated positive feedback mechanism. The Brazilian Journal of Infectious Diseases. 2017;**21**(1):19-26.

[89] Bateman RM, Sharpe MD, Singer M, Ellis CG. The effect of sepsis on the erythrocyte. Int J Mol Sci. 2017;**18**:1932. doi:10.3390/ijms18091932.

[90] Sadaka F, O'Brien J, Prakash S. Red cell distribution width and outcome in patients with septic shock. J Intensive Care Med. 2013;**28**(5):307-313.

[91] Fontana V, Spadaro S, Bond O, Cavicchi FZ, Annoni F, Donadello K, et al. No relationship between red blood cell distribution width and microcirculatory alterations in septic patients. Clin Hemorheol Microcirc. 2017;**66**(2):131-141.

[92] Huet O, Obata R, Aubron C, Spraul-Davit A, Charpentier J, Laplace C, et al. Plasma-induced endothelial oxidative stress is related to the severity of septic shock. Critical Care Medicine. 2007;**35**(3):821-826.

[93] Ekregbesi P, Shankar-Hari M, Bottomley C, Riley EM, Mooney JP. Relationship between Anaemia, Haemolysis, inflammation and Haem Oxygenase-1 at admission with sepsis: A pilot study. Sci Rep. 2018;**8**(1):11198.

[94] Uojima H, Onoue M, Hidaka H, Wada N, Tanaka Y, Inoue T, et al. A suspected case of Clostridium perfringens sepsis with intravascular hemolysis after transhepatic arterial chemoembolization: A case report. J Med Case Rep. 2019;**13**(1):125.

[95] Larsen R, Gozzelino R, Jeney V, Tokaji L, Bozza FA, Japiassu AM, et al. A Central Role for Free Heme in the Pathogenesis of Severe Sepsis. Science Translational Medicine. 2010;**2**(51): 51ra71.

[96] Lin G-L, McGinley JP, Drysdale SB, Pollard AJ. Epidemiology and immune pathogenesis of viral sepsis. Front Immunol. 2018;**9**:2147.

[97] Remy KE, Brakenridge SC, Francois B, Daix T, Deutschman CS, Monneret G, et al. Immunotherapies for COVID-19: Lessons learned from sepsis. Lancet Respir Med. 2020;**8**(10):946-949.

[98] Beltrán-García J, Osca-Verdegal R, Pallardó FV, Ferreres J, Rodríguez M, Mulet S, et al. Sepsis and coronavirus disease 2019: Common features and anti-inflammatory therapeutic approaches. Critical Care Medicine. 2020;**48**(12):1841-1844.

[99] Shindo Y, Fuchs AG, Davis CG, Eitas T, Unsinger J, Burnham C-AD, et al. Interleukin 7 immunotherapy improves host immunity and survival in a two-hit model of Pseudomonas aeruginosa pneumonia. J Leukoc Biol. 2017;**101**(2):543-554.

[100] Francois B, Jeannet R, Daix T, Walton AH, Shotwell MS, Unsinger J, et al. Interleukin-7 restores lymphocytes in septic shock: The IRIS-7 randomized clinical trial. JCI Insight. 2018;**3**(5):e98960.

[101] Unsinger J, Burnham C-AD, McDonough J, Morre M, Prakash PS, Caldwell CC, et al. Interleukin-7 ameliorates immune dysfunction and improves survival in a 2-hit model of fungal sepsis. The Journal of Infectious Diseases. 2012;**206**(4):606-616.

[102] Sterling D, Reithmeier RA, Casey JR. A transport metabolon. Functional interaction of carbonic anhydrase II and chloride/bicarbonate exchangers. J Biol Chem. 2001;**276**(51):47886-47894.

[103] Rivera-Santiago R, Harper SL, Sriswasdi S, Hembach P, Speicher DW. Full-length anion exchanger 1 structure and interactions with Ankyrin-1 determined by zero length crosslinking of erythrocyte membranes. Structure. 2017;**25**(1):132-145.

[104] Genetet S, Desrames A, Chouali Y, Ripoche P, Lopez C, Mouro-Chanteloup I. Stomatin modulates the activity of the anion exchanger 1 (AE1, SLC4A1). Sci Rep. 2017;7:46170.

[105] Hohensinner PJ, Mayer J, Kichbacher J, Kral-Pointner J, Thaler B, Kaun C, et al. Alternative activation of human macrophages enhances tissue factor expression and production of extracellular vesicles. Haematologica. 2021;**106**(2):454-463.

[106] Bernhardt I, Wesseling MC, Nguyen DB, Kaestner L. Red blood cells actively contribute to blood coagulation and thrombus formation. Erythrocyte. 2019. doi:10.5772/intechopen.86152.

[107] Steffen P, Jung A, Nguyen DB, Müller T, Bernhardt I, Kaestner L, et al. Stimulation of human red blood cells leads to Ca2+−mediated intercellular adhesion. Cell Calcium. 2011;**50**(1):54-61.

[108] Thangaraju K, Neerukonda SN, Katneni U, Buehler PW. Extracellular vesicles from red blood cells and their evolving roles in health, coagulopathy and therapy. Int J Mol Sci. 2020;**22**(1):153.

[109] Whelihan MF, Zachary V, Orfeo T, Mann KG. Prothrombin activation in blood coagulation: The erythrocyte contribution to thrombin generation. Blood. 2012;**120**(18):3837-3845.

[110] Koshiar RL, Somajo S, Norström E, Dahlbäck B. Erythrocyte-derived microparticles supporting activated protein C-mediated regulation of blood coagulation. PLoS One. 2014;**9**(8): e104200.

[111] Sluyter R. P2X and P2Y receptor signaling in red blood cells. Front Mol Biosci. 2015;**2**:60.

[112] Fagerberg S, Skals M, Leipziger J, Praetorius H. P2X receptor-dependent erythrocyte damage by α-Hemolysin from Escherichia coli triggers phagocytosis by THP-1 cells. Toxins. 2013;**5**(3):472-487.

[113] Lew VL, Tiffert T. On the mechanism of human red blood cell longevity: Roles of calcium, the sodium pump, PIEZO1, and Gardos channels. Front Physiol. 2017;**8**:977.

[114] Naeini MB, Bianconi V, Pirro M, Sahebkar A. The role of phosphatidylserine recognition receptors in multiple biological functions. Cell Mol Biol Lett. 2020;**25**(1):23.

[115] Burger et al. CD47 in Erythrocyte Ageing and Clearance - the Dutch Point of View. Transfus Med Hemother. 2012;**39**:348-352.

[116] Badior KE, Casey JR. Molecular mechanism for the red blood cell senescence clock. IUBMB Life. 2018;**70**(1):32-40.

[117] Sun Q, Zhao Y, Yang Y, Yang X, Li M, Xu X, et al. Loss of the clock protein PER2 shortens the erythrocyte life span in mice. Journal of Biological Chemistry. 2017;**292**(30):12679-12690.

[118] Repsold L, Joubert AM. Eryptosis: An Erythrocyte's suicidal type of cell death. BioMed Research International. 2018;**2018**:1-10.

[119] Föller M, Lang F. Ion transport in Eryptosis, the suicidal death of erythrocytes. Front Cell Dev Biol. 2020;**8**:597.

[120] Klei TRL, Meinderts SM, van den Berg TK, van Bruggen R. From the cradle to the grave: The role of macrophages in erythropoiesis and Erythrophagocytosis. Front Immunol. 2017;**8**:73.

[121] Montecinos L, Eskew JD, Smith A. What is next in this "age" of Heme-driven pathology and protection by Hemopexin? An update and links with iron. Pharmaceuticals. 2019;**12**(4):144.

[122] Belcher JD. Identification of a Heme activation site on the MD-2/TLR4 complex. Frontiers in Immunology. 2020;**11**:10.

[123] Bourne JH, Martina Colicchia, Ying Di, Eleyna Martin, Alexander Slater, Lubka T. Roumenina, et al. Heme induces human and mouse platelet activation through C-type-lectin-like receptor-2. Haematologica. 2020;**106**(2):626-629.

[124] Merle NS, Grunenwald A, Rajaratnam H, Gnemmi V, Frimat M, Figueres M-L, et al. Intravascular hemolysis activates complement via cell-free heme and heme-loaded microvesicles. JCI Insight. 2018;**3**(12): e96910.

[125] Cherry AD, Piantadosi CA. Regulation of mitochondrial biogenesis and its intersection with inflammatory responses. Antioxidants & Redox Signaling. 2015;**22**(12):965-976.

[126] Jamerson BD, Haryadi TH, Bohannon A. Glucose-6-phosphate dehydrogenase deficiency: An actionable risk factor for patients with COVID-19? Arch Med Res. 2020;**51**(7):743-744.

[127] Recht J, Ashley EA, White NJ. Use of primaquine and glucose-6-phosphate dehydrogenase deficiency testing: Divergent policies and practices in malaria endemic countries. PLOS Neglected Tropical Diseases. 2018;**12**(4):e0006230.

[128] Beauverd Y, Adam Y, Assouline B, Samii K. COVID-19 infection and treatment with hydroxychloroquine cause severe haemolysis crisis in a patient with glucose-6-phosphate dehydrogenase deficiency. Eur J Haematol. 2020. doi:10.1111/ejh.13432.

[129] Boulet C, Doerig CD, Carvalho TG. Manipulating Eryptosis of human red blood cells: A novel antimalarial strategy? Front Cell Infect Microbiol. 2018;**8**:419.

Chapter 3

Modulations in Oxidative Stress of Erythrocytes during Bacterial and Viral Infections

Vani Rajashekaraiah, Carl Hsieh and Masannagari Pallavi

Abstract

Oxidative stress (OS) occurs when the generation of free radicals and reactive oxygen species (ROS) overwhelms the antioxidant capacity. OS causes storage lesions which can be defined as a series of biochemical and biomechanical changes. Erythrocytes are constantly exposed to OS due to the presence of ROS, which are countered by the endogenous antioxidant system. Various irreversible changes that occur include fragmentation and aggregation of proteins and lipids. The changes in proteins, lipids and antioxidant capacity are used as OS biomarkers to assess the efficacy of the erythrocytes, post oxidative insult. Aging of erythrocytes is also associated with the changes in its physical, biochemical and physiological properties and OS causes its rapid aging. Bacterial and viral infections also cause OS which alters the erythrocytes' antioxidant capacity. These modulations in its microenvironment are both beneficial in terms of protection against invading microorganisms as well as harmful to the erythrocytes, causing damage to surrounding cells and tissues. Thus, OS biomarkers can be used to gain insights into the effects of bacterial and viral infections on the erythrocyte microenvironment.

Keywords: Erythrocytes, Young and Old Erythrocytes, Oxidative stress, Antioxidant capacity

1. Introduction

Oxidative stress is defined as an imbalance between oxidants and antioxidants leading to excessive levels of reactive oxygen species (ROS). Oxidative damage includes oxidative modification of cellular macromolecules, cell death by apoptosis or necrosis, as well as structural tissue damage. DNA, proteins and lipids, are the natural targets of oxidation [1].

In recent years, ROS have gained more attention, because of their central role in the progression of many inflammatory diseases [2]. The radical groups include hydroxyl radical ($OH^{\cdot}$), nitric oxide ($NO^{\cdot}$) and superoxide ($O_2^{\cdot-}$). Non-radical compounds can also be highly reactive, which includes peroxynitrite ($ONOO^-$), hydrogen peroxide (H_2O_2) and hypochlorous acid (HOCl) [3]. ROS can be described as oxygen free radicals and other non-radical oxygen derivatives involved in oxygen radical production [4], which are generated by the cells in most tissues and involved in normal cellular metabolism [5].

IntechOpen

ROS can react with DNA and cause damage to purines and pyrimidines [6], resulting in the formation of 8-Hydroxy-deoxyguanosine (8-OHdG) [7]. ROS also cause polypeptide chain fragmentation and covalent crosslinking that results in changes in its protein functional activity [8]. The covalent modifications of proteins induced by ROS or by reacting with secondary products of oxidative stress is termed as protein oxidation. These changes lead to many consequences such as, inhibition of enzyme activity, binding activity, aggregation, proteolysis, increased or decreased cell uptake, altered immunogenicity.

Protein oxidation serves as marker for determination of the levels of OS *in vitro*. There are many mechanisms that can induce protein oxidation as all the amino acyl side chains can be oxidatively modified. Cysteine and Methionine are the two amino acids that are most susceptible to oxidative attack due to the presence of sulfur atoms. Oxidation of cysteine leads to the formation of disulfide bonds, mixed disulfides and thiyl radicals, whereas modification of Methionine produces Methionine sulfoxide [8].

Lipids are important constituents of the lipid bilayer of the cellular membrane. Unsaturated fatty acids, which are easily oxidized, initiate the chain reactions, resulting in further oxidative damage. Lipids are susceptible to oxidation and reacts with molecular oxygen to form lipid peroxyl radicals which further oxidizes the neighboring lipids and propagates the oxidative damage [3]. Lipid peroxidation results in the changes of structural integrity and functioning of cell membranes. Lipid peroxidation markers such as malondialdehyde (MDA), 4-hydroxyl-2-nonenal (HNE), and isoprostane are used to evaluate oxidative damage.

Another category of substances called antioxidants exist in the cells and can effectively delay or inhibit ROS-induced oxidation. Antioxidants present in erythrocytes can be broadly classified into enzymatic and non-enzymatic antioxidants. There are three main groups of enzymatic antioxidants that play a significant role in protecting the cells from OS. Superoxide Dismutase (SOD) catalyzes the conversion of superoxides into hydrogen peroxide (H_2O_2) & oxygen (O_2). H_2O_2 is less toxic than compared to superoxides. SOD are metal containing enzymes that depend on a bound Mg, Cu or Zn for their antioxidant activity. There are 3 major families of SOD: Cu/Zn SOD, Fe/Mn SOD and Ni SOD [9–11]. Catalase (CAT) is the most common enzyme that is found in nearly all living organisms. It is found in the peroxisome of eukaryotic cells. It degrades H_2O_2 to H_2O & O_2. Hence it finishes the detoxification reaction started by SOD [12]. Glutathione peroxidase (GPx) contain Selenium, with peroxidase activity whose main activity is to protect the organism against OS. These enzymes like catalase degrade H_2O_2 to H_2O. They also reduce organic peroxides to their corresponding alcohol, thus provides another route for detoxification [13].

The extracellular endogenous antioxidants generally include the transition metal binding proteins i.e. ceruloplasmin, transferrin, hepatoglobin and albumin [14] and Vitamin C, α-tocopherol and Glutathione [15].

a. Glutathione (GSH), a major thiol antioxidant is a multifunctional intracellular antioxidant It is one of the most important cellular antioxidants due to its high concentration and its role in maintaining the redox state of the cell. GSH is a cysteine containing peptide and possesses antioxidant property due to the thiol group that serves as a reducing agent which can be reversibly oxidized and reduced. [16, 17].

b. Vitamin C, also known as L-ascorbic acid or ascorbate, is present naturally in the body and interconverts between each other depending on the pH. Ascorbic acid is well known reducing agent, thus serves as a good antioxidant [18]. Vit C is a water-soluble electron donor vitamin as it donates two of its electrons from the C-2 and C-3 double bond to act as antioxidants that result in the formation of semi dehydroascorbic acid, an intermediate free radical. This then reduces to a neutral ascorbate molecule [14]. In cells, ascorbic acid is maintained in its reduced form by reacting with glutathione, catalyzed by protein disulfide isomerase and glutaredoxins [17].

c. Vitamin E is a collective name that is given for a set of eight related tocopherols and tocotrienols that are fat soluble. α-tocopherol is the most commonly occurring natural antioxidant and has a phytyl chain that is attached to its chromanol nucleus [14]. It is a lipid soluble antioxidant and acts in the glutathione peroxidase pathway [19]. It protects the cell membranes against lipid peroxidation chain reaction by reacting with the lipid radicals produced due to OS. It removes the free radical intermediates, thus preventing the cascade and further damage [20].

1.1 Bacterial and viral and infections cause OS

Bacterial infections cause OS, which trigger ROS production leading to organ damage by altering the metabolic pathway [21].

1.1.1 Periodontitis- Fusobacterium nucleatum

Periodontitis is a common bacterial infection, caused *Fusobacterium nucleatum,* resulting in the destruction of teeth supporting tissues. Periodontitis is associated with overproduction of ROS by neutrophils. It is characterized by increased metabolites of lipid peroxidation, DNA damage and protein damage.

After pathogen stimulation, neutrophils produce $O_2^{\cdot-}$ via the metabolic pathway called "respiratory burst" catalyzed by NADPH oxidase during phagocytosis [22]. $O_2^{\cdot-}$can be released into phagosome and gets converted to different radical and non-radical derivatives, such as hydrogen peroxide (H_2O_2), hypochlorous acid (HOCl), hydroxyl radical ($OH^{\cdot}$) and singlet oxygen ($_1O_2$). *In vitro* studies have shown that not only neutrophils, but also other phagocytes and cells of periodontal tissues such as monocytes, gingival fibroblast and periodontal ligament cells exhibit enhanced ROS production upon stimulation by periodontal pathogens and their components [23–25]. This was evident in the results of lipid peroxidation and protein oxidation products. Higher levels of TBARS, MDA and 8-isoprostanes were found in blood plasma, erythrocytes, gingival crevicular fluid (GCF) [26], saliva, GCF [27–30] respectively in periodontitis patients compared to healthy controls. Higher levels of protein carbonyl groups (PC) were found in GCF, saliva and serum of periodontitis patients [31–34] and 8-Hydroxy-deoxyguanosine (8-OHdG) in GCF and saliva of periodontitis patients [35–44].

Variations in antioxidant enzymes were also observed in many studies. SOD, CAT, GPx and glutathione reductase activities decreased in saliva of periodontitis patients [36, 45]. However, SOD and CAT in plasma, erythrocytes and gingival tissues were elevated, whereas activities of non-enzymatic antioxidants (vitamins E, vitamin C, and reduced glutathione) decreased in periodontitis [26]. Periodontitis is associated with decreased Total antioxidant capacity (TAC) [46–52].

1.1.2 Tuberculosis- mycobacterium tuberculosis

Mycobacteria initiate infection at oxygen rich lung microenvironments, generating oxidative radicals. These toxic radicals kill the pathogens by causing disintegration of bacterial cell membrane, DNA damage, deactivation of metabolic enzymes or proteins [53–56]. After invasion into the host, mycobacteria induce NADPH oxidase 2 (NOX2) expression to generate superoxide radicals ($O_2^{\cdot -}$), which are then converted to more toxic hydrogen peroxide (H_2O_2) by superoxide dismutase (SOD) and subsequently reduced to water and molecular oxygen by catalase [57, 58]. NADPH oxidase 2 (NOX2) is the key enzyme responsible for the cellular ROS production by using superoxide radicals (O_2-) as precursor molecule [59]. Alterations in regulatory components of NOX2 results in generation of phagocytic oxidative stress and phagocytic burst to eliminate enclosed pathogen [58].

However, pathogenic mycobacteria can inhibit oxidative stress mechanisms by modulation of cell signaling mechanisms, up-regulation of antioxidant enzymes and redox buffering systems [59–62].

1.1.3 Pneumococcal meningitis- streptococcus pneumoniae

Pneumococcal meningitis is a life-threatening disease characterized by acute infection affecting the pia mater, arachnoid, and subarachnoid spaces [63]. *Streptococcus pneumoniae* crosses the blood–brain barrier (BBB) and disrupts the intraepithelial tight junctions. Host polymorphonuclear leukocytes produce nitric oxide, superoxide radicals, and hydrogen peroxide in response to bacterial infection. $O_2 - \bullet$ and NO• can lead to the formation of peroxynitrite (ONOO), a strong oxidant [64–66]. $ONOO^-$ can damage neurons and glial cells by lipid peroxidation and cell membrane destabilization, resulting in DNA disintegration and subsequent poly (ADP-ribose) polymerase (PARP) activation. Elevated 4-HNE and MDA levels are found in bacterial meningitis patients [67]. Thus, ROS/ RNS can be considered key players of immune activation, blood–brain barrier disruption, vascular failure, neuronal injury, and cochlear damage during pneumococcal meningitis.

1.1.4 Gastritis/gastric cancer- Helicobacter pylori

Helicobacter pylori is the causative pathogen for human gastritis or gastric cancer, which is characterized with inflammation and ulceration of the stomach and duodenum. Gastric cancer arises from oxidative stress and environmental toxins, which increase DNA mutation rates [68]. The possible sources of ROS/ RNS in *H. pylori* infected stomach, include neutrophils, vascular endothelial cells and gastric mucosal cells. Neutrophils are believed to be the main source of ROS/ RNS [69] and their production is catalyzed by NADPH oxidase on the cell membrane [59]. These highly reactive ROS (HOCl and $^{\cdot}$OH) are used by the phagocyte to kill pathogenic bacteria. *H. pylori* infected gastric mucosal phagocytes produce greater amounts of ROS, which is believed to be the major cause of gastric mucosal damage.

1.1.5 HIV and Hepatitis

Oxidative Stress has always played a major pathogenic role in HIV and hepatitis infections. HIV causes decrements in glutathione (GSH), cystine, vitamin C and SOD levels, and increments in lipid peroxidation [70–73]. A decline in the antioxidant capacity represents a weakened immune system, thus requiring more antioxidants to maintain normal functionality [74]. Hepatitis, similar to HIV, also increases the lipid peroxidation (malondialdehyde (MDA) and 4-hydroxynonenal (HNE)) and activity of caspases, whereas reduces zinc [75–77] (**Table 1**).

OS Markers	Bacterial diseases	Viral diseases	References
TBARS	Periodontitis- Increased Tuberculosis- Increased Sepsis- Increased		[26] [78] [79]
MDA	Periodontitis- Increased Meningitis- Increased	HBV- Increased HCV- Increased HIV- Increased JEV- Increased RSV- Increased	[26] [80] [81]
8-IP	Periodontitis- Increased		[26]
4-HNE	Meningitis- Increased		[80]
PC	Periodontitis- Increased Sepsis- Increased		[31] [79]
8-OHdG	Periodontitis- Increased	HCV- Increased	[35] [81]
NO	Tuberculosis- Decreased Sepsis- Increased	RSV- Increased	[78] [82] [81]
NT	Meningitis- Increased Sepsis- Increased		[80] [82]
SOD	Periodontitis- Decreased Gastritis- Decreased	HBV- Decreased HIV- Decreased JEV- Increased DENV-Decreased RSV- Decreased	[36, 45] [83] [81]
CAT	Periodontitis- Decreased Tuberculosis- Decreased	Paramyxovirus-Decreased DENV-Decreased RSV- Decreased	[36, 45] [78] [81]
GPx	Periodontitis- Decreased Gastritis- Decreased	HBV- Decreased Paramyxovirus-Decreased DENV-Decreased RSV- Decreased	[36, 45] [83] [81]
GSH	Sepsis- Decreased	HIV- Decreased JEV- Increased Influenza virus-Decreased	[82] [81]
GR	Periodontitis- Decreased		[36, 45]
GST		Paramyxovirus-Decreased RSV- Decreased	[81]
Vit-C	Sepsis- Increased Meningitis- Decreased	Influenza virus-Decreased HIV- Decreased Covid-19- Decreased	[79] [80] [81] [84]
Vit-E		Covid-19- Decreased Influenza virus-Decreased	[81] [84]
β-Carotene		Covid-19- Decreased	[84]
Bilurubin, UA, FRAP, TRAP	Sepsis- Increased		[79]

HBV- Hepatitis B virus; HCV- Hepatitis C virus; HIV- Human immunodeficiency virus; JEV- Japanese encephalitis virus; DENV- Dengue virus; RSV- Respiratory syncytial virus; Covid-19-Coronavirus Disease 2019.
TBARS- Thiobarbituric acid reactive substances; MDA- Malondialdehyde; 8-IP- 8-isoprostanes; 4-HNE- 4-hydroxynonenal; PC- Protein Carbonyls; 8-OHdG- 8-Hydroxy-deoxyguanosine; NO- nitric oxide; NT- Nitrotyrosine; SOD- Superoxide dismutase; CAT- Catalase; GPx- Glutathione Peroxidase; GSH- Glutathione; GR- Glutathione Reductase, GST- Glutathione S-Transferase; UA- Uric acid; Vit-C- Vitamin C; Vit-E- Vitamin E; FRAP- ferric reducing antioxidant power; TRAP- Total-trapping radical antioxidant potential.

Table 1.
Oxidative stress (OS) in bacterial and viral diseases.

2. OS has dual role in diseases

OS offers protection against invading microorganisms, and on the other hand can cause damage to cells/tissues.

Erythrocytes being heme rich, provides the invading bacteria a rich source of iron for its metabolism. Bacteria have evolved mechanisms to scavenge the iron through hemolytic toxins and heme scavenging systems [85–87]. Erythrocytes counteract the bacteria through the production of ROS. The α and β sub units of hemoglobin possess high affinity binding sites for lipopolysaccharides (LPS), and leads to macrophage cytokine production and enhances the macrophage binding to LPS. Hemoglobin in μM concentrations have shown to inhibit yeast and bacterial growth through the production of ROS [88–90].

The protection conferred by hemoglobin against invading organisms have a detrimental effect in case of pathological states. Elevations in free hemoglobin is associated with increased mortality. Globin is associated with the protective properties, whereas heme triggers the proinflammatory response. During endotoxemia, the protective effects of hemoglobin is attributed to globin scavenging free heme, which has the property of activating a host of proinflammatory proteins and ROS generation. It has also been shown to increase the transcription of proinflammatory genes by 100-fold [91–93].

3. Modulations in erythrocytes due to OS

3.1 ROS cascade

The ROS cascade in erythrocytes begins with the autooxidation of hemoglobin (Hb) into methemoglobin (MetHb). Oxidation of MetHb results in the formation of sulfhemoglobin (SulfHb) along with superoxide anion (O_2). Erythrocytes have an innate antioxidant system that detoxifies the cells. The superoxide generated is detoxified by superoxide dismutase into H_2O and H_2O_2, which is further detoxified by catalase and glutathione peroxidase (with the help of glutathione) into H_2O and O_2. Erythrocytes also contain non-enzymatic antioxidants such as glutathione, ascorbic acid (Vit C), and α-tocopherol (Vit E). This antioxidant mechanism helps the erythrocyte's survival in an oxygen-rich environment. Glutathione redox system: reduced glutathione (GSH), glutathione disulfide (GSSG), glutathione reductase (GR), glutathione peroxidase plays an important role in inactivating the ROS [94, 95].

3.2 Erythrocyte aging

The lifespan of erythrocytes *in vivo* is around 120 days. About 1% of the erythrocytes are cleared or phagocytized from circulation every day in humans. The membrane of erythrocytes comprises of proteins (50%), lipids (40%), and carbohydrates (10%). Hb comprises about 95% of the total cytoplasmic proteins. The membrane-associated proteins include band 3 (anion exchanger), band 4.1, spectrin, ankyrin, and glycophorin C which are responsible for maintaining the structure of the cell [94].

As the reticulocyte ages and transforms into erythrocyte, various changes occur in its membrane, composition, appearance and catalytic functions [96]. Aging of erythrocytes is associated with the changes in physical, biochemical and physiological properties. Thus, aged cells are more prone to be trapped and ultimately destroyed during microcirculation [97]. During aging, a decrease in the cell volume and hemoglobin is observed. The old erythrocytes also increase in density as they bind to autologous IgG (immunoglobulin G), that serves as an initiator for the removal of senescent erythrocytes. The binding of antibodies is also associated and triggered by changes to the anion exchanger, Band 3. There is a decline in sulphate transport with age, thus hampering the binding of ankyrin to Glyceraldehyde 3-phosphate dehydrogenase (GAPDH), a key enzyme in glycolysis [98]. The N- and C- terminal regions of Band 3 are conformationally changed during aging that results in the formation of neoantigens, which serve as a senesce marker. It is also observed that damaged hemoglobin (Hb) bind to band 3 resulting in cluster formations [99–102]. There is also an increase in the amount of glycated Hb.

Aging erythrocytes also lose water, 2,3-BPG, ATP, proteins, Hb and vesicles that cause the cell volume and surface charge to decrease. There is also loss of some of the surface materials such as sialic acids that alter the structure and function of the membrane. These sialic acids are 90% *N*-acetylneuraminic acid (NANA) which accords the electrical charge to the cells [103]. These senescent RBCs also expose membrane phosphatidylserine. An increase in erythrocyte OS causes accelerated aging, resulting in decreased function and survival [97].

The study of Igbokwe *et al.*, 1994 reported the effects of *Trypanosoma brucei* infection in mice. They concluded that a lowered ability to prevent lipid peroxidation in infected mice may increase aging of erythrocytes [104].

4. Bacterial and viral infections causing variations in erythrocytes

Erythrocytes were assumed to only function as innate oxygen carriers. However, recent studies have shown them to be important in modulating the innate immune response [105–107]. Mammalian erythrocytes, unlike the erythrocytes of birds, amphibians and fishes, are enucleate and lack major cell organelles. The organelles of the latter modulate the immune response through production of cytokine like factors, upregulating viral response genes and pathogen sequestering through phagocytosis. Mammalian erythrocytes on the other hand modulate the innate immune response through generation of reactive oxygen species (ROS) to promote inflammatory and autoimmune response against invading microorganisms [108].

Minasyan, 2014 highlighted the role of erythrocytes in conferring bacterial immunity, which is comparable to phagocytic leukocytes as: (i) they are more numerous in number; (ii) fend off microorganisms repeatedly without injury; and (iii) resistant to infection. Erythrocytes have a longer lifespan when compared with leukocytes as well as being produced at a faster rate. The cytosol of erythrocytes is unfavorable to parasitic organisms such as chlamidiae, mycoplasmas, rickettsiae, viruses, etc. Erythrocytes elevate to the primary line of defense against bacterial infections when: (i) there is a presence of massive microbial load; (ii) ineffective recruitment of phagocytes; (iii) faster proliferation and spread of the microorganisms than the phagocytes' capacity; and (iv) ineffectiveness of the phagocytes against the invading microorganisms [109].

4.1 Bacterial infections

Sepsis is one of the most recognized life-threatening dysfunctions that is caused due to infections and is the leading cause for mortality in non-cardiac ICUs (intensive care units) around the world. Sepsis may be caused by gram-positive, gram-negative and poly microbial infection. It results in elevated OS caused due to inflammatory response, which alters erythrocytes leading to phagocytosis by macrophages and polymorphonuclear leukocytes (PMN). This activation of the macrophages and PMN results in a positive feedback mechanism, as the modified erythrocytes trigger its continuous activation with the generation of ROS. This mechanism may further lead to septic shock, even if the blood-derived bacteria have been cleared off the system [110].

Sepsis develops when bacteria in the bloodstream survive oxidation on the surface of erythrocytes [109]. The changes in erythrocytes can be caused due to several reasons, one of them being OS, and these interactions seem to be interconnected. The levels of antioxidants and oxidants are inversely proportional in septic patients. Decrements in Vitamin E, ascorbate, β-carotene and retinol, while increments in lipid peroxidation were observed. Antioxidant supplementation improved the outcome of patients. Thus, erythrocytes can be model cells for the management of sepsis/septic shock to improve the outcome of patients [111]. Larsen *et al.*, 2010, have reported that one of the major causes for sepsis is the increasing levels of free heme released due to hemolysis. It can be sequestered and cleared off using hemopexin, as its administration reduced tissue damage and lethality [112].

4.2 Viral infections

Studies have shown that viruses cause cell death by generating OS within infected cells [113–115]. The influenza virus and parmovirus activates monocytes to generate ROS *in vitro* [113]. The influenza virus also possesses hemagglutinin glycoprotein on its surface, which helps the virus in binding to cells rich in sialic acids like erythrocytes. Influenza carries hemagglutinin in its inactive form (HOA), which is activated by proteases by cleaving HOA into hemagglutinin 1 and hemagglutinin 2 (HA1 and HA2). HOA can be activated by ROS resulting in a non-infectious virus getting converted into an infectious one [116, 117]. Thus, an increase in ROS levels is beneficial for the invading influenza virus.

COVID-19, caused by severe acute respiratory syndrome coronavirus-2 (SARS-CoV-2) was believed to interact with Hb, facilitating the removal of heme, which was proposed by Liu and Li as likely pathway for the loss of function of hemoglobin (Hb) and the accumulation of free heme resulting in elevated OS. De Martino *et al.*, 2020, reported that COVID-19 did not exhibit any hemolytic anemia. The levels of Hb, bilirubin, lactate dehydrogenase, iron, ferritin and haptoglobin in COVID-19 patients were similar to those with acute respiratory distress syndrome (ARDS) not infected with COVID-19, suggesting that the oxygen delivery impairment was not due to red cell hemolysis and removal of iron from heme. [118].

4.3 Bacterial v/s viral infections

Trefler et al., 2014 compared the patterns of OS in bacterial origin community acquired pneumonia [BCAP] and 2009 A/H1N1 virus community acquired pneumonia [VCAP], revealing the distinct responses between bacterial and viral infections. Erythrocyte GR activity was significantly higher in patients with VCAP in respect of BCAP patients. Lower TBARS levels were observed in VCAP patients in comparison to BCAP, suggesting an increase of antioxidant activity related to the redox glutathione system. GR, GSSG, GSSG/GSH and GPx levels were more elevated in patients with viral pneumonia. A higher antioxidant activity in patients with 2009 A/H1N1 viral pneumonia was observed (**Figure 1**) [119].

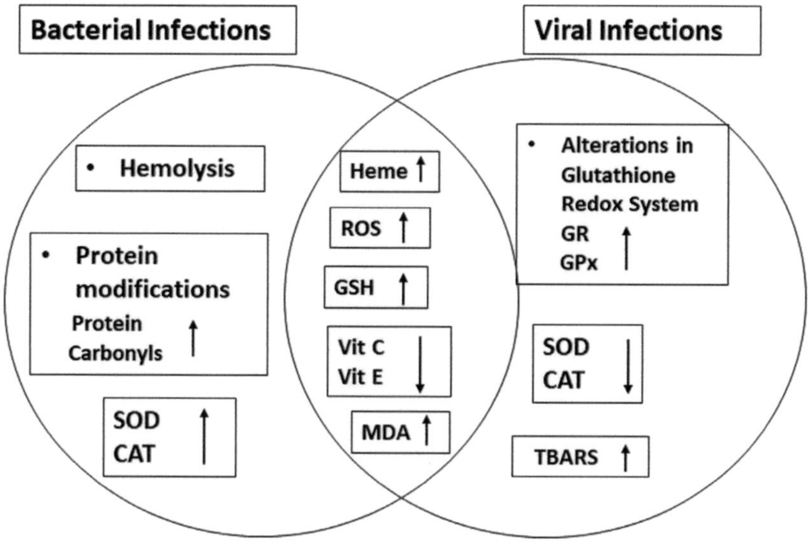

Figure 1.
Oxidative modifications in Erythrocytes during Bacterial and Viral infections [109, 112, 118–120]. SOD – Superoxide dismutase; CAT – Catalase; ROS – Reactive oxygen species; GSH – Glutathione; Vit C – Vitamin C; Vit E – Vitamin E; MDA – Malondialdehyde; GR – Glutathione reductase; GPx – Glutathione peroxidase; TBARS – Thiobarbituric acid reactive substances.

5. Conclusion

ROS levels increase rapidly leading to lipid peroxidation and protein oxidation in erythrocytes during infections. Generally, there is a decline in the antioxidant capacity of erythrocytes. Nevertheless, some microbes evade their destruction by altering the antioxidant enzymes of erythrocytes. Thus, OS biomarkers can be used to gain insights into the effects of bacterial and viral infections on the erythrocyte microenvironment. Therefore, erythrocytes act as good indicators and can be promising candidates as peripheral biomarkers during bacterial and viral infections.

Acknowledgements

The authors acknowledge Dr. Leela Iyengar, and JAIN (Deemed-to-be University) for their support.

Conflict of interest

The authors have no conflict of interest to disclose.

Author details

Vani Rajashekaraiah*, Carl Hsieh and Masannagari Pallavi
Department of Biotechnology, School of Sciences, JAIN (Deemed-to-be University), Bengaluru, India

*Address all correspondence to: tiwari.vani@gmail.com; vani.rs@jainuniversity.ac.in

References

[1] Poulsen HE. Oxidative DNA modifications. Experimental and Toxicologic Pathology. 2005;57:161-169.

[2] Mittal M, Siddiqui MR, Tran K, Reddy SP, Malik AB. Reactive oxygen species in inflammation and tissue injury. Antioxidants and Redox Signalling. 2014;20:1126-1167.

[3] Lykkesfeldt J, Svendsen O. Oxidants and antioxidants in disease: Oxidative stress in farm animals, The Veterinary Journal. 2007;173:502-511.

[4] Lushchak VI. Free radicals, reactive oxygen species, oxidative stress and its classification. Chemico-Biological Interactions. 2014;224:164-175.

[5] Sies H. Oxidative stress: Oxidants and antioxidants. Experimental Physiology. 1997;82:291-295.

[6] Halliwell, B. Why and how should we measure oxidative DNA damage in nutritional studies? How far have we come? American Journal of Clinical Nutrition. 2000;72:1082-1087.

[7] Chapple IL, Matthews JB. (2007). The role of reactive oxygen and antioxidant species in periodontal tissue destruction. Periodontology 2007;43:160-232.

[8] Shacter E. (2000). Quantification and significance of protein oxidation in biological samples. Drug Metabolism Reviews. 2000;32:307-326.

[9] Tainer JA, Getzoff ED, Beem KM, Richardson JS, Richardson DC. Determination and analysis of the 2 A-structure of copper, zinc superoxide dismutase. Journal of Molecular Biology. 1982;160:181-217.

[10] Borgstahl GE, Parge HE, Hickey MJ, Beyer WF, Hallewell RA, Tainer JA. The structure of human mitochondria manganese superoxide dismutase reveals a novel tetrameric interface of two 4-helix bundles. Cell. 1992;71:107-118.

[11] Barondeau DP, Kassmann CJ, Bruns CK, Tainer JA, Getzoff ED. Nickel superoxide dismutase structure and mechanism. Biochemistry. 2004;43:8038-8047.

[12] Chelikani P, Fita L, Loewen PC. Diversity of structures and properties among catalases. Cellular and Molecular Life Sciences. 2004;61:192-208.

[13] Otto EPP, Ladenstein R, Wendel A. The refined structure of the selenoenzyme Glutathione Peroxidase at 0.2-nm resolution. European Journal of Biochemistry. 1983;133:51-69.

[14] Chakraborty P, Kumar S, Dutta D, Gupta V. Role of antioxidants in common health diseases. Research Journal of Pharmacy and Technology. 2009;2:238-244.

[15] Machefer G, Groussard C, Rannou-Bekono F, Zouhal H, Faure H, Vincent S, Cillard J, Gratas-Delamarche A. Extreme running competition decreases blood antioxidant defense capacity. Journal of the American College of Nutrition. 2004;23:358-364.

[16] Rahman K. Studies on free radicals, antioxidants, and co-factors. Clinical Interventions in Aging. 2007;2:219-236.

[17] Hamid AA, Aiyelaagbe OO, Usman LA, Ameen OM, Lawal A. Antioxidants: Its medicinal and pharmacological applications. African Journal of Pure and Applied Chemistry. 2010;4:142-151.

[18] McGregor G P, Biesalski H K. Rationale and impact of vitamin C in clinical nutrition. Current Opinion in

Clinical Nutrition & Metabolic Care. 2006;9:697-703.

[19] Wefers H, Sies H. The protection by ascorbate and glutathione against microsomal lipid peroxidation is dependent on vitamin E. European Journal of Biochemistry. 1988;174:2:353-357.

[20] Herrera E, Barbas C. Vitamin E: Action, metabolism and perspectives. Journal of Physiology and Biochemistry. 2001;57:43-56.

[21] Ivanov AV, Bartosch B, Isaguliants MG. Oxidative stress in infection and consequent disease. Oxidative Medicine and Cellular Longevity. 2017;3496043:1-3.

[22] Chapple IL, Matthews JB. The role of reactive oxygen and antioxidant species in periodontal tissue destruction. Periodontology 2007;43:160-232

[23] Bullon P, Cordero MD, Quiles JL, Morillo JM, del Carmen Ramirez-Tortosa M, Battino M. Mitochondrial dysfunction promoted by Porphyromonas gingivalis lipopolysaccharide as a possible link between cardiovascular disease and periodontitis. Free Radical Biology and Medicine. 2011;50:1336-1343.

[24] Chang MC, Tsai YL, Chen YW, Chan CP, Huang CF, Lan WC, Lin CC, Lan WH, Jeng JH. Butyrate induces reactive oxygen species production and affects cell cycle progression in human gingival fibroblasts. Journal of Periodontal Research. 2013;48:66-73.

[25] Gölz L, Memmert S, Rath-Deschner B, Jäger A, Appel T, Baumgarten G, Götz W, Frede S. LPS from P. gingivalis and hypoxia increases oxidative stress in periodontal ligament fibroblasts and contributes to periodontitis. Mediators of Inflammation. 2014;2014.

[26] Panjamurthy K, Manoharan S, Ramachandran CR. Lipid peroxidation and antioxidant status in patients with periodontitis. Cellular and Molecular Biology Letters. 2005;10: 255-264.

[27] Tsai CC, Chen HS, Chen SL, Ho YP, Ho KY, Wu YM, Hung CC. Lipid peroxidation: A possible role in the induction and progression of chronic periodontitis. Journal of Periodontal Research. 2005;40:378-384.

[28] Wei D, Zhang XL, Wang YZ, Yang CX, Chen G. Lipid peroxidation levels, total oxidant status and superoxide dismutase in serum, saliva and gingival crevicular fluid in chronic periodontitis patients before and after periodontal therapy. Australian Dental Journal. 2010;55:70-78.

[29] Tonguç MÖ, Öztürk Ö, Sütçü R, Ceyhan BM, Kılınç G, Sönmez Y, Yetkin Ay Z, Şahin Ü, Baltacıoğlu E, Kırzıoğlu FY. The impact of smoking status on antioxidant enzyme activity and malondialdehyde levels in chronic periodontitis. Journal of Periodontology. 2011;82:1320-1328.

[30] Pradeep AR, Rao NS, Bajaj P, Agarwal E. 8-Isoprostane: A lipid peroxidation product in gingival crevicular fluid in healthy, gingivitis and chronic periodontitis subjects. Archives of Oral Biology. 2013;58:500-504.

[31] Sculley DV, Langley-Evans SC. Periodontal disease is associated with lower antioxidant capacity in whole saliva and evidence of increased protein oxidation. Clinical Science. 2003;105:167-172.

[32] Baltacioglu E, Akalin FA, Alver A, Deger O, Karabulut E. Protein carbonyl levels in serum and gingival crevicular fluid in patients with chronic periodontitis. Archives of Oral Biology. 2008;53:716-722.

[33] Pradeep AR, Ramchandra prasad MV, Bajaj P, Rao NS, Agarwal E. Protein carbonyl: An oxidative stress marker in gingival crevicular fluid in healthy, gingivitis, and chronic periodontitis subjects. Contemporary Clinical Dentistry. 2013;4:27-31.

[34] Nguyen TT, Ngo LQ, Promsudthi A, Surarit R. Salivary oxidative stress biomarkers in chronic periodontitis and acute coronary syndrome. Clinical Oral Investigations. 2017;21:2345-2353.

[35] Takane M, Sugano N, Iwasaki H, Iwano Y, Shimizu N, Ito K. New biomarker evidence of oxidative DNA damage in whole saliva from clinically healthy and periodontally diseased individuals. Journal of Periodontology. 2002;73:551-554.

[36] Canakci CF, Cicek Y., Yildirim A, Sezer U, Canakci V. (2009). Increased levels of 8-hydroxydeoxyguanosine and malondialdehyde and its relationship with antioxidant enzymes in saliva of periodontitis patients. European journal of dentistry. 2009;3:100-106.

[37] Su H, Gornitsky M, Velly AM, Yu H, Benarroch M, Schipper HM. Salivary DNA, lipid, and protein oxidation in nonsmokers with periodontal disease. Free Radical Biology and Medicine. 2009; 46:914-921.

[38] Sezer U, Erciyas K, Üstün K, Pehlivan Y, Ziya Şenyurt S, Aksoy N, Tarakçıoğlu M, Taysı S, Onat AM. Effect of chronic periodontitis on oxidative status in patients with rheumatoid arthritis. Journal of Periodontology. 2013;84:785-792.

[39] Dede FO, Ozden FO, Avci B. 8-hydroxy-deoxyguanosine levels in gingival crevicular fluid and saliva in patients with chronic periodontitis after initial periodontal treatment. Journal of Periodontology. 2013;84:821-828.

[40] Hendek MK, Erdemir EO, Kisa U, Ozcan G. Effectof initial periodontal therapy on oxidative stress markers in gingival crevicular fluid, saliva, and serum in smokers and non-smokers with chronic periodontitis. Journal of Periodontology. 2015;86:273-282.

[41] Kurgan Ş, Önder C, Altıngöz SM, Bağış N, Uyanık M, Serdar MA, Kantarcı A. High sensitivity detection of salivary 8-hydroxy deoxyguanosine levels in patients with chronic periodontitis. Journal of Periodontal Research. 2015;50:766-774.

[42] Zamora-Perez AL, Ortiz-García YM, Lazalde-Ramos BP, Guerrero-Velázquez C, Gómez-Meda BC, Ramírez-Aguilar MÁ, Zúñiga-González GM. Increased micronuclei and nuclear abnormalities in buccal mucosa and oxidative damage in saliva from patients with chronic and aggressive periodontal diseases. Journal of Periodontal Research. 2015;50:28-36.

[43] Shin MS, Shin HS, Ahn YB, Kim HD. Association between periodontitis and salivary 8-hydroxydeoxyguanosine among Korean rural adults. Community Dentistry Oral Epidemiology. 2016;44:381-389.

[44] Önder C, Kurgan Ş, Altıngöz SM, Bağış N, Uyanık M, Serdar MA, Kantarcı A, Günhan M. Impact of non-surgical periodontal therapy on saliva and serum levels of markers of oxidative stress. Clinical Oral Investigations. 2017;21:1961-1969.

[45] Trivedi S, Lal N, Mahdi AA, Mittal M, Singh B, Pandey S. Evaluation of antioxidant enzymes activity and malondialdehyde levels in patients with chronic periodontitis and diabetes mellitus. Journal of Periodontology. 2014;85:713-720.

[46] Brock GR, Butterworth CJ, Matthews JB, Chapple IL. Local and systemic total antioxidant capacity in

periodontitis and health. Journal of Clinical Periodontology. 2004;31: 515-521.

[47] Chapple IL, Brock GR, Milward MR, Ling N, Matthews JB. Compromised GCF total antioxidant capacity in periodontitis: Cause or effect? Journal of Clinical Periodontology. 2007;34:103-110.

[48] Guentsch A, Preshaw PM, Bremer-Streck S, Klinger G, Glockmann E, Sigusch BW. Lipid peroxidation and antioxidant activity in saliva of periodontitis patients: Effect of smoking and periodontal treatment. Clinical Oral Investigations. 2008;12:345-352.

[49] Baltacıoğlu E, Yuva P, Aydın G, Alver A, Kahraman C, Karabulut E, Akalın FA. Lipid peroxidation levels and total oxidant/antioxidant status in serum and saliva from patients with chronic and aggressive periodontitis. Oxidative stress index: A new biomarker for periodontal disease?. Journal of Periodontology. 2014;85:1432-1441.

[50] Baser U, Gamsiz-Isik H, Cifcibasi E, Ademoglu E, Yalcin F. Plasma and salivary total antioxidant capacity in healthy controls compared with aggressive and chronic periodontitis patients. Saudi Medical Journal. 2015;36:856-861.

[51] Zhang T, Andrukhov O, Haririan H, Müller-Kern M, Liu S, Liu Z, Rausch-Fan X. Total antioxidant capacity and total oxidant status in saliva of periodontitis patients in relation to bacterial load. Frontiers in Cellular and Infection Microbiology. 2016;5:97.

[52] Ahmadi-Motamayel F, Goodarzi MT, Jamshidi Z, Kebriaei R. Evaluation of salivary and serum antioxidant and oxidative stress statuses in patients with chronic periodontitis: A case-control study. Frontiers in Physiology. 2017;8:189.

[53] Gupta S, Chatterji D. Stress responses in mycobacteria. IUBMB Life 2005;57:149-159.

[54] Dos Vultos T, Mestre O, Tonjum T, Gicquel B. DNA repair in mycobacterium tuberculosis revisited. FEMS Microbiology Reviews. 2009;33:471-487.

[55] Nambi S, Long JE, Mishra BB, Baker R, Murphy KC, Olive AJ, Nguyen HP, Shaffer SA, Sassetti CM. The oxidative stress network of mycobacterium tuberculosis reveals coordination between radical detoxification systems. Cell Host and Microbe. 2015;17:829-837.

[56] Tyagi P, Dharmaraja AT, Bhaskar A, Chakrapani H, Singh A. Mycobacterium tuberculosis has diminished capacity to counteract redox stress induced by elevated levels of endogenous superoxide. Free Radical Biology and Medicine. 2015;84:344-354.

[57] Voskuil MI, Bartek IL, Visconti K, Schoolnik GK. The response of mycobacterium tuberculosis to reactive oxygen and nitrogen species. Frontiers in Microbiology. 2011;2:105.

[58] Dan Dunn J, Alvarez LA, Zhang X, Soldati T. Reactive oxygen species and mitochondria: A nexus of cellular homeostasis. Redox Biology. 2015;6:472-485.

[59] Lambeth JD. NOX enzymes and the biology of reactive oxygen. Nature Reviews Immunology. 2004;4:181-189

[60] Mohanty S, Jagannathan L, Ganguli G, Padhi A, Roy D, Alaridah N, Saha P, Nongthomba U, Godaly G, Gopal RK, Banerjee S. A mycobacterial phosphoribosyltransferase promotes bacillary survival by inhibiting oxidative stress and autophagy pathways in

macrophages and zebrafish. Journal of Biological Chemistry. 2015;290:13321-13343.

[61] Mohanty S, Dal Molin M, Ganguli G, Padhi A, Jena P, Selchow P, Sengupta S, Meuli M, Sander P, Sonawane A. Mycobacterium tuberculosis EsxO (Rv2346c) promotes bacillary survival by inducing oxidative stress mediated genomic instability in macrophages. Tuberculosis. 2016;96:44-57.

[62] Chao WC, Yen CL, Hsieh CY, Huang YF, Tseng YL, Nigrovic PA, Shieh CC. Mycobacterial infection induces higher interleukin-1β and dysregulated lung inflammation in mice with defective leukocyte NADPH oxidase. PloS One. 2017;12:e0189453.

[63] Mook-Kanamori BB, Geldhoff M, van der Poll T, van de Beek D. Pathogenesis and pathophysiology of pneumococcal meningitis. Clinical Microbiology Reviews. 2011;24:557-591..

[64] Klein M, Koedel U, Pfister HW. Oxidative stress in pneumococcal meningitis: A future target for adjunctive therapy?. Progress in Neurobiology. 2006;80:269-280.

[65] Mitchell TJ. Virulence factors and the pathogenesis of disease caused by Streptococcus pneumoniae. Research in Microbiology. 2000;151:413-419.

[66] Braun JS, Sublett JE, Freyer D, Mitchell TJ, Cleveland JL, Tuomanen EI, Weber JR. Pneumococcal pneumolysin and H 2 O 2 mediate brain cell apoptosis during meningitis. The Journal of clinical investigation. 2002;109:19-27.

[67] Kastenbauer S, Koedel U, Becker BF, Pfister HW. Oxidative stress in bacterial meningitis in humans. Neurology. 2002;58:186-191.

[68] Correa P, Cuello C, Duque E, Burbano LC, Garcia FT, Bolanos O, Brown C, Haenszel W. Gastric cancer in Colombia. III. Natural history of precursor lesions. Journal of the National Cancer Institute. 1976;57:1027-1035.

[69] Naito Y, Yoshikawa T. Molecular and cellular mechanisms involved in helicobacter pylori-induced inflammation and oxidative stress. Free Radical Biology and Medicine. 2002;33:323-336.

[70] Reshi ML, Su YC, Hong JR. RNA viruses: ROS-mediated cell death. International journal of cell biology. 2014;2014.

[71] Malvy DJ, Richard MJ, Arnaud J, Favier A, Amédée-Manesme O. Relationship of plasma malondialdehyde, vitamin E and antioxidant micronutrients to human immunodeficiency virus-1 seropositivity. Clinica Chimica Acta. 1994;224:89-94.

[72] Fuchs J, Emerit I, Levy A, Cernajvski L, Schöfer H, Milbradt R. Clastogenic factors in plasma of HIV-1 infected patients. Free Radical Biology and Medicine. 1995;19:843-848.

[73] Peterhans E. Reactive oxygen species and nitric oxide in viral diseases. Biological Trace Element Research. 1997;56:107-116.

[74] Sprietsma JE. Zinc-controlled Th1/Th2 switch significantly determines development of diseases. Medical Hypotheses. 1997;49:1-14.

[75] Boya P, de la Peña A, Beloqui O, Larrea E, Conchillo M, Castelruiz Y, Civeira MP, Prieto J. Antioxidant status and glutathione metabolism in peripheral blood mononuclear cells from patients with chronic hepatitis C. Journal of Hepatology. 1999;31:808-814.

[76] De Maria N, Colantonl A, Fagiuoli S, Liu GJ, Rogers BK, Farinati F,

Van Thiel DH, Floyd RA. Association between reactive oxygen species and disease activity in chronic hepatitis C. Free Radical Biology and Medicine. 1996;21:291-295.

[77] Bianchi GP, Marchesini G, Brizi M, Rossi B, Forlani G, Boni P, Melchionda N, Thomaseth K, Pacini G. Nutritional effects of oral zinc supplementation in cirrhosis. Nutrition Research. 2000;20:1079-1089.

[78] Rajopadhye HS, Mukherjee RS, Chowdhary SA, Dandekar PS. Oxidative stress markers in tuberculosis and HIV/TB co-infection. Journal of Clinical and Diagnostic Research. 2017;11:24-28.

[79] Andresen M, Regueira T, Bruhn A, Perez D, Strobel P, Dougnac A, Marshall G, Leighton F. Lipo peroxidation and protein oxidative damage exhibit different kinetics during septic shock. Mediators of Inflammation. 2008;2008:168652.

[80] Kastenbauer S, Koedel U, Becker BF, Pfister HW. Oxidative stress in bacterial meningitis in humans. American Academy of Neurology 2002;58: 186-191.

[81] Camini CF, da Silva Caetano CC, Almeida TL, de Brito Magalhaes LC. Implications of oxidative stress on viral pathogenesis. Archives of Virology. 2017;162:907-917.

[82] Bavunoglu I, Genc H, Konukoglu D, Cicekci H, Sozer V, Gelisgen R, Uzun H oxidative stress parameters and inflammatory and immune mediators as markers of the severity of sepsis. Journal of Infection in Developing Countries. 2016;10:1045-1052.

[83] Augusto AC, Miguel F, Mendonç S, Pedrazzoli Jr J, Gurgueir SA. Oxidative stress expression status associated to helicobacter pylori virulence in gastric diseases. Clinical Biochemistry. 2007;40:615-622.

[84] Pincemai J, Cavalier E, Charlier C, Cheramy–Bien J P, Brevers E, Courtois A, Fadeur M, Meziane S, Goff C L, Misset B, Albert A, Defraigne J O, Rousseau A F. Oxidative Stress status in COVID-19 patients hospitalized in Intensive Care Unit for Severe Pneumonia. A Pilot Study. Antioxidants 2021;10:257.

[85] Diaz-Ochoa V, Jellbauer S, Klaus S, Raffatellu M. Transition metal ions at the crossroads of mucosal immunity and microbial pathogenesis. Frontiers in Cellular and Infection Microbiology. 2014;4:2.

[86] Deriu E, Liu JZ, Pezeshki M, Edwards RA, Ochoa RJ, Contreras H, Libby SJ, Fang FC, Raffatellu M. Probiotic bacteria reduce salmonella typhimurium intestinal colonization by competing for iron. Cell Host and Microbe. 2013;14:26-37.

[87] Skaar EP, Humayun M, Bae T, DeBord KL, Schneewind O. Iron-source preference of Staphylococcus aureus infections. Science. 2004;305:1626-1628.

[88] Bahl N, Du R, Winarsih I, Ho B, Tucker-Kellogg L, Tidor B, Ding JL. Delineation of lipopolysaccharide (LPS)-binding sites on hemoglobin: From in silico predictions to biophysical characterization. Journal of Biological Chemistry. 2011;286:37793-37803.

[89] Bodet C, Chandad F, Grenier D. Hemoglobin and LPS act in synergy to amplify the inflammatory response. Journal of Dental Research. 2007;86:878-882.

[90] Liepke C, Baxmann S, Heine C, Breithaupt N, Ständker L, Forssmann WG. Human hemoglobin-derived peptides exhibit antimicrobial activity: A class of host defense

peptides. Journal of Chromatography B. 2003;791:345-356.

[91] Hao K, Hanawa H, Ding L, Ota Y, Yoshida K, Toba K, Ogura M, Ito H, Kodama M, Aizawa Y. Free heme is a danger signal inducing expression of proinflammatory proteins in cultured cells derived from normal rat hearts. Molecular Immunology. 2011;48:1191-1202.

[92] Yang H, Wang H, Bernik TR, Ivanova S, Wang H, Ulloa L, Roth J, Eaton JW, Tracey KJ. Globin attenuates the innate immune response to endotoxin. Shock. 2002;17:485-490.

[93] Su D, Roth RI, Yoshida M, Levin J. Hemoglobin increases mortality from bacterial endotoxin. Infection and Immunity. 1997;65:1258-1266.

[94] Çimen MB. Free radical metabolism in human erythrocytes. Clinica Chimica Acta. 2008;390:1-11.

[95] Tatsuro Y, Prudent M, D'Alessandro A. Red blood cell storage lesion: Causes and potential clinical consequences. Blood Transfusion. 2019;17:27-52.

[96] Ahn J, Johnstone RM. Synthesis of the transferrin receptor in peripheral sheep reticulocytes: evidence for incomplete oligosaccharide processing. Red Blood Cell Aging. Springer, Boston, MA. 1991:3-13

[97] Nash GB, Wyard SJ. Changes in surface area and volume measured by micropipette aspiration for erythrocytes ageing in vivo. Biorheology. 1980;17:479-484

[98] Bosman GJCGM, Werre JM, Willekens FLA, Novotný VMJ. Erythrocyte ageing in vivo and in vitro: Structural aspects and implications for transfusion. Transfusion Medicine. 2008;18:335-347.

[99] Bosman GJCGM, Willekens FLA, Werre JM. Erythrocyte aging: A more than superficial resemblance to apoptosis?. Cellular and Physiological Biochemistry. 2005;16:1-8.

[100] Belo L, Rebelo I, Castro EM, Catarino C, Pereira-Leite L, Quintanilha A, Santos-Silva A. Band 3 as a marker of erythrocyte changes in pregnancy. European Journal of Haematology. 2002;69:145-151.

[101] Ciccoli L, Rossi V, Leoncini S, Signorini C, Blanco-Garcia J, Aldinucci C, Buonocore G, Comporti M. Iron release, superoxide production and binding of autologous IgG to band 3 dimers in newborn and adult erythrocytes exposed to hypoxia and hypoxia-reoxygenation. Biochimica et Biophysica Acta (BBA)-General Subjects. 2004;1672:203-213.

[102] Rossi V, Leoncini S, Signorini C, Buonocore G, Paffetti P, Tanganelli D, Ciccoli L, Comporti M. Oxidative stress and autologous immunoglobulin G binding to band 3 dimers in newborn erythrocytes. Free Radical Biology and Medicine. 2006;40:907-915.

[103] Huang YX, Wu ZJ, Mehrishi J, Huang BT, Chen XY, Zheng XJ, Liu WJ, Luo M. Human red blood cell aging: Correlative changes in surface charge and cell properties. Journal of Cellular and Molecular Medicine. 2011;15:2634-2642.

[104] Igbokwe IO, Esievo KA, Saror DI, Obagaiye OK. Increased susceptibility of erythrocytes to in vitro peroxidation in acute Trypanosoma brucei infection of mice. Veterinary Parasitology. 1994;55:279-286.

[105] Hotz MJ, Qing D, Shashaty MG, Zhang P, Faust H, Sondheimer N, Rivella S, Worthen GS, Mangalmurti NS. Red blood cells homeostatically bind mitochondrial

DNA through TLR9 to maintain quiescence and to prevent lung injury. American Journal of Respiratory and Critical Care Medicine. 2018;197:470-480.

[106] Neote K, Darbonne W, Ogez J, Horuk R, Schall TJ. Identification of a promiscuous inflammatory peptide receptor on the surface of red blood cells. Journal of Biological Chemistry. 1993;268:12247-12249.

[107] Darbonne WC, Rice GC, Mohler MA, Apple T, Hébert CA, Valente AJ, Baker J. Red blood cells are a sink for interleukin 8, a leukocyte chemotaxin. The Journal of clinical investigation. 1991;88:1362-1369.

[108] Anderson HL, Brodsky IE, Mangalmurti NS. The evolving erythrocyte: Red blood cells as modulators of innate immunity. The Journal of Immunology. 2018;201:1343-1351.

[109] Minasyan H. Erythrocyte and blood antibacterial defense. European Journal of Microbiology and Immunology. 2014;4:138-143.

[110] de Oliveira YPA, Pontes-de-Carvalho LC, Couto RD, Noronha-Dutra AA. Oxidative stress in sepsis. Possible production of free radicals through an erythrocyte-mediated positive feedback mechanism. The Brazilian Journal of Infectious Diseases. 2017;21:19-26.

[111] Moisă E, Negoiță S, Corneci D. Understanding red blood cell rheology in sepsis and its role in clinical practice. From biomolecular aspects to possible therapeutic interventions. Central European Journal of Clinical Research. 2018;1:40-58.

[112] Larsen R, Gozzelino R, Jeney V, Tokaji L, Bozza FA, Japiassú AM, Bonaparte D, Cavalcante MM, Chora Â, Ferreira A, Marguti I. A central role for free heme in the pathogenesis of severe sepsis. Science Translational Medicine. 2010;2:51ra71.

[113] Peterhans E, Grob M, Burge TH, Zanoni R. Virus-induced formation of reactive oxygen intermediates in phagocytic cells. Free radical research communications. 1987 Jan 1;3(1-5):39-46.

[114] Vierucci A, De Martino M, Graziani E, Rossi ME, London WT, Blumberg BS. A mechanism for liver cell injury in viral hepatitis: Effects of hepatitis B virus on neutrophil function in vitro and in children with chronic active hepatitis. Pediatric Research. 1983;17:814-820.

[115] Müller F. Reactive oxygen intermediates and human immunodeficiency virus (HIV) infection. Free Radical Biology and Medicine. 1992;13:651-657.

[116] Belding ME, Klebanoff SJ, Ray CG. Peroxidase-mediated virucidal systems. Science. 1970;167:195-196.

[117] Russell RJ, Kerry PS, Stevens DJ, Steinhauer DA, Martin SR, Gamblin SJ, Skehel JJ. Structure of influenza hemagglutinin in complex with an inhibitor of membrane fusion. Proceedings of the National Academy of Sciences. 2008;105: 17736-17741.

[118] DeMartino AW, Rose JJ, Amdahl MB, Dent MR, Shah FA, Bain W, McVerry BJ, Kitsios GD, Tejero J, Gladwin MT. No evidence of hemoglobin damage by SARS-CoV-2 infection. Haematologica. 2020;105:2769-2773.

[119] Trefler S, Rodriguez A, Martin-Loeches I, Sanchez V, Marina J, Llaurado M, Romeu M, Diaze E, Nogues R, Giralt M. Oxidative stress in

immunocompetent patients with severe community-acquired pneumonia. A pilot study. Medicinia Intensiva. 2014;38:73-82.

[120] Minasyan H. Sepsis and septic shock: Pathogenesis and treatment perspectives. Journal of Critical Care. 2017;40:229-242.

Section 2

Functional Biomarkers

Chapter 4

Early Predictive Biomarkers for Hypertension Using Human Fetal Astrocytes

Fahmida Abdi, Ann M. Simpson, Sara Lal and Kaneez Fatima Shad

Abstract

Hypertension is a major risk factor for cardiovascular and cerebrovascular diseases, causing high numbers of deaths and /or disabilities worldwide. Previous studies have reported numerous biomolecules, such as, triglycerides and fibrinogen as biomarkers of hypertension (HTN), but none of these biomolecules could be considered as 'true' predictive biomarkers as they were produced after the establishment of HTN. Therefore, there is an urgent need for identifying and monitoring molecules that are linked to early pre-HTN stages, that is, prior to the onset of HTN. Astrocytes are the most abundant cells in the nervous system and through their long processes, astrocytes can communicate with both neuronal and non-neuronal cells such as endothelial cells lining blood vessels. Thus, any biochemical changes in astrocytes will affect both blood vessels and neurons. We are using human fetal astrocytes (HFAs) to investigate the molecules which may possibly act as early predictive biomarkers for hypertension. Astrocytic processes are mostly supported by the intermediate filaments, an example is the glial fibrillary acidic protein (GFAP) which is a type III intermediate filament. Elevated GFAP levels are being considered as a marker of astroglial injury, indicating the conversion of non-reactive (A2) into reactive (A1) astrocytes. Our initial immunohistochemistry studies using anti-GFAP antibodies on astrocytes from spontaneous hypertensive rats (SHRs) and their normal counter parts (WKY) rats showed a similar profile to that of reactive (A1) and non-reactive (A2) HFAs, respectively. Numerous studies point to a significant role of calcium ion channel proteins in hypertension, and calcium channel blockers such as Amlodipine (Norvasc) Diltiazem (Cardizem) are commonly used as antihypertensive drugs. By using liquid chromatography–tandem mass spectrometry (LC–MS/MS) we observed that reactive (A1) astrocytes, contain more calcium-activated proteins such as calpain, calpastatin, cathepsin and mitogen activated protein kinase (MAPK) as compare to normal (A2) HFAs, suggesting their possible link to the future onset of HTN. Hence these proteins could be considered as potential early predictive biomarkers of HTN.

Keywords: Hypertension, spontaneous hypertensive rats (SHRs), WKY rats, Human fetal astrocytes (HFAs), Reactive (A1) and non-reactive (A2) HFAs, glial fibrillary acidic protein (GFAP), Ca2+ dependent proteins, LC–MS/MS technique

1. Introduction

Scientific studies have established that prevalence of hypertension (HTN) needs to be reduced in order to control cardiovascular and cerebrovascular diseases [1]. At present, many laboratories around the world are exploring a plethora of biomarkers, as early indicators of HTN [2, 3], i.e., triglycerides, C-reactive protein, fibrinogen, serum albumin, uric acid, homocysteine and intracellular adhesion molecule-1(ICAM-1) [4]. All these biomarkers for hypertension are produced as a result of chronic disease-related comorbidities such as atherosclerosis, Type 2 diabetes and renal failure, therefore, they cannot be considered as 'true' predictive biomarkers for hypertension. Hence, this study is focusing on glial fibrillary acidic protein (GFAP), and a group of calcium-dependent proteases, such as calpain, calpastatin, cathepsin and mitogen activated protein kinase (MAPK), known to be strongly associated with HTN; hence could be potential early biomarkers for HTN.

Similarly, an endogenous chemical, N-Methyl-D-Aspartic acid (NMDA), a type of glutamate is also documented as linked with HTN [5]. NMDA is a major excitatory neurotransmitter in the central nervous system, consisting of two types of receptors namely synaptic and extrasynaptic. Research indicates that the activation of the synaptic NMDA receptors is neuroprotective; whereas, the stimulation of extrasynaptic NMDA receptors (NMDARs) promotes cell death [6] and resultant HTN, which can be regulated by NMDA antagonists, such as 1-amino cyclo propane carboxylic acid (ACPC) [7]. NMDA type glutamate receptors are not only present in the neuronal cells but also in the non-neuronal cells such as astrocytes [8]. Recent studies have demonstrated a significant role of astrocytes in regulating blood flow due to the elevation in intracellular calcium (Ca2+) [9], which plays a significant role in the regulation of blood pressure.

Astrocytes are present between blood vessels and neurons and are responsible for changes in the arterial blood pressure [10]. A decline in the cerebral blood supply activates the astrocytes to release a chemical signal to the nearby neurons that raises blood pressure, restoring blood flow and oxygen supply to the brain [10]. Thus, astrocytes perform a balancing role between brain perfusion and neuronal activities by mobilising their internal calcium [11], which in turn triggers the release of chemical transmitters such as glutamate [12]. Consequently, there is a calcium-dependent bidirectional signalling pathway between astrocytes and neurons [13], which opens up the possibility of astrocytic involvement in the modulation of calcium-dependent molecules such as calpain, calpastatin, cathepsin and MAPK, which can potentially be considered as direct predictive biomarkers of HTN.

Certain behavioural conditions, such as, stress also elevates blood pressure. Under stressful conditions, excessive release of corticotropin-releasing hormone (CRH) activates NMDA receptors, resulting in an influx of Ca2+ molecules, which enhance the activity of m-calpain [14]. Calpain is one of the major calcium-dependent proteolytic enzymes with various isoforms such as μ-calpain and m-calpain, which are activated by the synaptic and extra-synaptic NMDA receptors respectively [15, 16]. Studies have indicated that the activation of μ-calpain is important for cell-survival, whereas the stimulation of m-calpain initiates toxic effects and cell death [6], due to its interaction with NMDAR [17]. Activation of the type of NMDA receptor defines the communication from the synapse to the nucleus [18].

Stimulation of synaptic NMDAR phosphorylates the intermediate filament proteins by extracellular signal regulated kinases 1 and 2 (ERK1/2) [19]. However, the activation of extrasynaptic NMDAR fails to phosphorylate and translocate intermediate filament proteins into the nucleus. The phosphorylated or non-phosphorylated state of intermediate filament proteins determine whether it promotes cell survival or induces cell death. These diverse functions that require the regulation of gene and synapse-to-nucleus communications are controlled by nuclear calcium signalling [20]. Overexpression of these proteins induces neurodegeneration, whereas, its suppression has the opposite effect, therefore, the molecules that keep the balance may act as therapeutic targets for HTN.

To date, most proteomic research, targeted at discovering biomarkers, has failed to incorporate adequate biomarker validation studies in independent sample sets. These are necessary steps in the translation of potential biomarkers into clinical practice. Recent development of sophisticated mass spectrometry-based quantitation of multiple proteins has enabled the validation of candidate biomarkers in different biological materials, such as astrocytes.

The literature indicates that reactive astrocytes (A1) and their intermediate filaments contain greater concentrations of glial fibrillary acidic protein (GFAP) [21], compared to normal astrocytes (A2). The purpose of this study was, to identify and quantify the differences between the proteins of interest, present in the reactive and the normal human foetal astrocytes (HFAs), using immunocytochemistry and single-pot, solid-phase-enhanced sample-preparation (SP3) [22] proteomic techniques. We aim to detect the immunocytochemical differences between the A1 and A2 HFAs using GFAP antibodies and to identify a panel of protein biomarkers in A1 and A2 HFAs for the prediction of hypertension.

2. Materials and methods

2.1 Ethical approval

HFAs were obtained from the Biobank of Macquarie University, after approval from the UTS Human Research Ethics Committee (ETH17–1883).

2.2 Protocol for the cell culture of primary HFAs

After getting a material transfer agreement (MTA # 17/979) between the two universities, the HFAs were tested for mycoplasma contamination and were found to be negative. The HFAs were then cultured in T75 flasks using Roswell Park Memorial Institute (RPMI 1640) media containing 10% heat-inactivated foetal calf serum (FCS) (**Table 1**), at 37°C, in a 5% CO_2 humidified incubator. Freshly prepared media was used to feed the cells every five days.

The HFAs were seeded at $4x10^7$ cells/T75-cm^2 in tissue culture flasks containing RPMI (4.5 g/L glucose, L-glutamine, and 25 mM HEPES buffer), plus 10% FCS at 37°C, and incubated in a 5% CO_2 humidified incubator.

For re-seeding, HFAs were detached from the flask by using 4 ml of trypsin, incubated for 3 minutes at 37°C, in a 5% CO_2 humidified incubator. After incubation with trypsin, the cells were transferred to a 15 ml falcon tube and were centrifuged at 1600 g rpm, for 3 minutes. Trypsin was removed by washing three times with phosphate buffered saline (PBS). The cells were then resuspended in 2 ml of fresh media by gentle mixing and 1 ml of this cell solution was added to each T75 flask with 14 ml RPMI, labelled as A1 and A2 for reactive and normal astrocytes, respectively.

2.3 Conversion of normal HFAs (A2) into reactive HFAs (A1)

After achieving 95 to 100% confluency and three days before the proteomics experiment, the HFAs were incubated in the RPMI media with 1% FCS instead of 10% FCS, to avoid false positive proteomic results due to the proteins present in the 10% FCS. Twenty-four hours before the pellet formation for SP3 [22] protocol,

S. N.	Chemical	Catalogue number	Company	Amount used	Quantity T75 flask
1	RPMI 1640	11875	Thermo Fisher Scientific	45 ml	15 ml of complete media
2	Foetal Calf Serum	A3160401	Thermo Fisher Scientific	5 ml	
3	Trypsin	9035-81-8	Sigma-Aldrich		4 ml
4	Phosphate-buffered saline (PBS)	10010023	Thermo Fisher Scientific		3 ml
5	Adenosine 5- triphosphate disodium salt hydrate (ATP)	34369–07-8	Sigma-Aldrich		100 μl
6	Paraformaldehyde	30525–89-4	Sigma-Aldrich	4% paraformaldehyde in PBS, with 7.4 pH	1 ml

Table 1.
Chemicals used for the tissue culture experiments.

the cells were returned to RPMI media +10% FCS in both A1 and A2 flasks. Then, 100 μl of 1 mM adenosine triphosphate (ATP) was added to the A1 flask to convert the normal HFAs into reactive HFAs [23]. After 24 hours of incubation at 37°C, in a 5% CO_2 humidified incubator, the cells were washed three times with PBS. The cells were subsequently detached from both flasks by adding 4 ml of trypsin and were incubated at 37°C in a humidified incubator, for 3 minutes. After incubation, the cells were transferred into two falcon tubes (15 ml tubes, labelled as A1 and A2) and were centrifuged at 1600 g rpm for 3 minutes. Trypsin was removed from the tubes by washing three times with 3 ml PBS. The cell samples from both tubes were transferred into 2 ml Eppendorf tubes and were microfuged for 1 minute, to remove PBS. Subsequently, the Eppendorf tubes containing cell pellets were snap frozen in liquid nitrogen and stored at -80°C for the proteomics experiment, using SP3 protocol.

2.4 Immunocytochemistry of HFAs using anti-GFAP stain

Six poly-L-lysine coated coverslips, three labelled as A1 and three as A2 were placed in a six well plate, containing 2 ml of fresh RPMI media with 10% FCS. The plates were incubated at 37°C, in a 5% CO2 humidified incubator to grow until ~70% confluency was achieved.

The cells in three of the six wells, were treated with 1 mM ATP to convert them into reactive (A1) astrocytes, whereas the other three coverslips with the HFAs (A2) were not treated with ATP. After 24 hours of incubation all the cells were washed three times with 0.1 M PBS, fixative (4% paraformaldehyde in PBS, 7.4 pH) was then added for 30 minutes, at room temperature. After fixation, both types of HFAs were gently washed three times with PBS and were ready for permeabilization and blocking of the unspecific binding of the proteins. To improve the permeabilization of the antibodies, the HFAs were incubated with 100 μl of 0.1% Triton X 100, for 30 minutes, at 37°C, in a 5% CO2 humidified incubator. The cells were gently washed three times with PBS, before being treated with the blocking solution (100 μl of 5% goat serum) and then the cells were incubated for 30 minutes.

The cells were incubated with a primary anti-GFAP antibody (1:100 dilution, ab7260 from ABCAM) at 4°C for 18 hours. Then, the HFAs were washed three times with PBS, and were incubated with the secondary antibody (1:100, ab150077 with AF488, ABCAM) in 1% BSA for 1 hour at the room temperature in the dark.

2.4.1 Nuclear staining and mounting

The HFAs were rinsed with PBS and were incubated with 1 µg/ml of nuclear stain [4, 6-diamidino-2-phenylindole (DAPI)] for 10 minutes followed by three rinses with PBS. Excess PBS was carefully wiped around the coverslips and the cells were mounted on the glass slides (75 mm x 25 mm and 1 mm thickness), which were clearly labelled with the type of cell and stain. Then a drop of mounting medium, Fluoroshield (ABCAM) was placed in the middle of the slide. The coverslip with the stained cells was carefully lifted using forceps and was placed in an inverted position on top of the slide, air bubbles were removed, and the cells were sealed with a clear nail polish to prevent drying and movement under the microscope. The cells were stored in the dark, at +4°C till image analysis was performed using the NIKON A1 confocal microscope.

2.5 Protein extraction and digestion using single-pot, solid-phase-enhanced sample-preparation (SP3) protocol

The proteins of interest were extracted from A1 and A2 HFAs using the SP3 protocol, before performing mass spectrometric analysis. All the chemicals and equipment used for the preparation of lysis buffer and for protein extraction and digestion using SP3 protocol are shown in **Table 2** and **Figure 1**.

Chemical name	Company	Cat. No.
Bovine serum albumin (BSA)	Sigma	00A2153
Sodium dodecyl sulfate (SDS)	Bio-Rad	1610302
Triton X-100	Sigma	T8787
Nonidet-P40 (NP-40)	Merck-Millipore	492016-100ML
Tween 20	Sigma	P1379
Deoxycholate	Sigma	3970
Sodium chloride (NaCl)	Sigma	S7653
Glycerol	Sigma	G5516
HEPES sodium salt	Sigma	H3375
Iodoacetamide	Bio-Rad	1632109
Dithiothreitol (DTT)	Bio-Rad	1610611
Ammonium bicarbonate	Sigma	A6141
Trypsin + rLysC mix	Promega	V5073
1.5-mL Safe-Lock tubes	Eppendorf	22363204
Tris-2-carboxyethyl phosphine	Sigma	51805–45-9
Sera-Mag Speed Beads	GE Healthcare	45152105050250
Magnetic rack, Magnetic Sphere	Promega	Z5342

Table 2.
Reagents and equipment used for SP3 protocol.

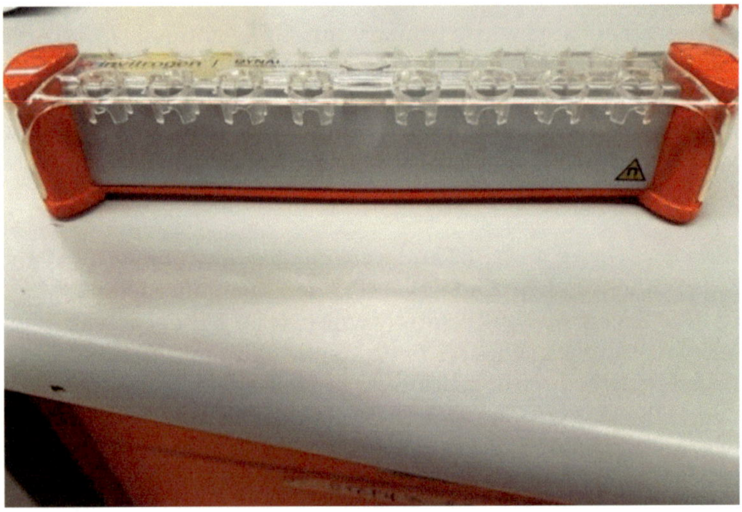

Figure 1.
The magnetic rack used for placing the sample-vials containing magnetic beads. When vials are placed on the magnetic rack, the beads start to stick to the wall of the vial. This process allows the protein molecules to attach safely to the beads, so unwanted fluid could be easily discarded.

2.5.1 Preparation of reconstitution (lysis) buffer for SP3

Reconstitution buffer was freshly prepared by mixing 100 μl of each of the following chemicals, i.e., 10% SDS, 10% Triton x-100, 10% NP-40, 10% Deoxycholate and 10% Glycerol. In addition, 10 μl of 5 M NaCl, 20 μl of HEPES (pH 8.0) and 470 μl of water was added to the mixture.

2.5.2 Extraction of the proteins from A1 and A2 HFA samples

Eppendorf tubes containing pellets of A1 and A2 astrocytes were removed from −80°C and were resuspended in 50 μl of lysis buffer. To facilitate the extraction process, the tubes were heated to 90°C on the heating block for 10 minutes. To reduce the number of disulfide bonds, 1 μl Tris 2-carboxyethyl phosphine (TCEP) was added to the samples, followed by 1 μl of acrylamide monomer to alkylate the mixture. After an hour of incubation at room temperature, the alkylation reaction was stopped by adding 1 μl of dithiothreitol (DTT). Before extracting the proteins from both types of HFAs, a standard curve was constructed using an ampule of 2 mg/ml bovine serum albumin (BSA).

2.5.3 Construction of the standard curve

The standard curve was constructed using a 96 well microplate, and bovine serum albumin (BSA) was added at a concentration ranging from 0 to 2000 μg/ml diluted with double distilled water. Similarly, A1 and A2 sample solutions were also diluted with double distilled water at a ratio of 1:10 & 1:20. Working solution of BCA was prepared by adding reagent B (cupric sulfate) with reagent A (cuprous sulfate), at a ratio of 1:50. The plates were prepared according to **Table 3**. Nine wells of the plate were filled with 25 μl of BSA solution ranging from 0 to 2000 μg/ml along with the blank and two dilutions of A1 and A2 samples. Then 200 μl of working solution was added in each of the fourteen wells. The plates were covered with parafilm and were incubated at 37°C for 30 minutes, and the plates were read on the Tecan Infinite 200 Pro, spectrophotometer, at a wavelength of 562 nm, using Magellan software.

S.N.	Vial volume of diluent (μl)	Volume and source of BSA (μl)	BSA concentration (μg/ml)
A	0	300 of Stock	2000
B	125	375 of Stock	1500
C	325	325 of Stock	1000
D	175	175 of vial B dilution	750
E	325	325 of vial C dilution	500
F	325	325 of vial E dilution	250
G	325	325 of vial F dilution	125
H	400	100 of vial G dilution	25
I	400	0	Blank
Samples			
A1(1:10)	9	1	38
A1(1:20)	19	1	26
A2(1:10)	9	1	22
A2(1:20)	19	1	18

Table 3.
Dilution scheme protocol for standard in vials (work-range, 25–2000 μg/mL).

Data obtained was used to plot a standard curve and the concentration of the proteins in the A1 and A2 astrocyte samples were determined from **Figure 2**.

2.5.4 Reconstitution of the proteins from A1 and A2 samples

The cell-solution was diluted to a final volume of 48 μl by adding double distilled water and 2 μl of hydrophilic magnetic beads to the samples, giving a total volume of 50 μl of protein solution. Proteins were bonded with the beads by adding 50 μl of 100% ethanol to the above mixture and vortexed for 5 minutes. Hydrophilic magnetic beads, were coated with carboxylate functional groups to capture the proteins, making it easier to wash the cells in 100% ethanol. Surfactant was removed by rinsing the cells three times with 80% ethanol and the supernatant with unbound

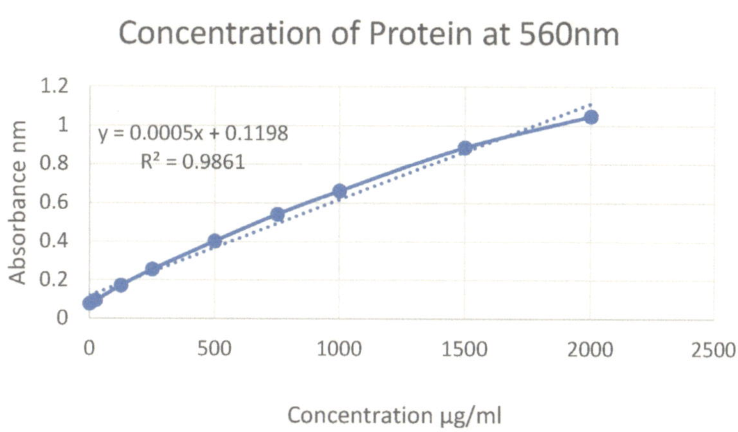

Figure 2.
BCA (Bicinchoninic acid) protein assay was used to plot standard curve for quantitation of the protein concentrations in the A1 and A2 astrocytes.

compounds was discarded in a waste container. Proteins were extracted in aqueous solution by adding 100 µl of 200 mM ammonium bicarbonate to the protein bonded beads. Trypsinisation of proteins was achieved by incubating the samples overnight with 1 µl of trypsin at 37°C.

2.5.5 STop and go extraction (STAGE) tips protocol

Further desalting and cleaning of the peptides was performed by an extended protocol called STop and Go Extraction (STAGE) tips protocol [22]. The first step was to prepare Styrenedivinylbenzene-reverse phase sulfonated (SDB-RPS) for the process of solid phase extraction. Briefly, a large sized SDB-RPS disk was placed in a covered petri dish and from that a 47 mm disk core was used in this protocol as a solid phase extraction sorbent. These sorbents have a high affinity for polar organic compounds, and hence reduce the extraction time and solvent, used for analytical preparation of aqueous environmental samples.

2.5.6 Construction of STAGE tip and preparing of collection tubes

STAGE tip (**Figure 3**) was prepared by using a flat-ended needle, to push the 47 mm SDB-RPS core carefully inside a 2 ml pipette-tip, keeping it 2–3 mm above its end. A hole was made in the middle of the lid of a 2 ml Eppendorf tube using a rotary cutting tool. The pipette tip with SDB-RPS disk was inserted into the Eppendorf tube through the hole. Double distilled water was used to make 10% v/v trifluoroacetic acid (TFA), then 90% of isopropanol was added to make 1% TFA and this was diluted 1:10 with water. Fresh elution solvent was made by mixing 700 µl of Acetonitrile, 71 µl of NH_4OH stock and 229 µl of water.

2.5.7 Desalting and washing the samples

The sample was acidified by adding 10 µl of 10% TFA and was centrifuged for 5 minutes at maximum speed. Using one STAGE tip, the sample was pipetted into the top of the tip and centrifuged at 5000 g rpm for 1 minute. The tubes were repeatedly centrifuged until all the liquid passed through the STAGE tip. To wash the sample, 60 µl of 1% TFA was pipetted into the STAGE tip and centrifuged at 5000 g rpm for 2 minutes. Tubes were centrifuged three times to clear all the salts and contaminants from the bound peptides and the liquid in the collection tube was discarded.

2.5.8 Setting up STAGE tip in autovial insert and elution of peptides

For each sample, a separate autovial was used and a yellow spacer was placed between the STAGE tip and the collection tube to hold the STAGE tip in the center of the autovial. Elution solvent was prepared by mixing 30 ml of 1 M Ammonium Hydroxide with 70 ml Acetonitrile. To complete the reaction, 60 µl of elution solvent was pipetted into the STAGE tip and was microfuged for a few seconds to drag the liquid onto the disk, which was incubated on the bench for 10 minutes. The STAGE tip was placed in the insert of the autovial and was centrifuged at 5000 g rpm for 5 minutes. This step was repeated three times to allow all the fluid to pass through the disk into the autovial insert containing all the peptides. For removing all the liquid, the STAGE tip was discarded and the collection tubes with autovial inserts containing peptide samples were placed into the Speed Vac (Thermo Speed Vac Concentrator DNA 120), to vacuum-centrifuge. Then, the autovial inserts were removed from the Speed Vac and the springs were attached back to the bottom of the inserts, before placing them into the autovials. Samples

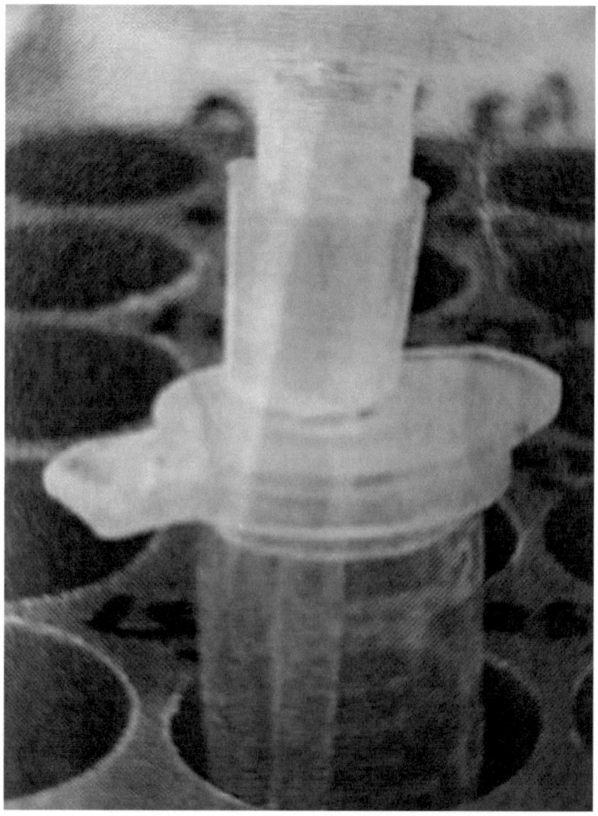

Figure 3.
Shows a 2 ml pipette-tip (STAGE tip) inserted through a hole drilled into the top of a 2 ml collection tube. The pipette tip with SDB-RPS disk was inserted into the Eppendorf tube through the hole. Cell solution was pipetted from the top into tip so peptides could react with the SDB-RPS disk for 10 minutes.

were prepared by adding 25 µl of MS Loading Solvent-A (2% Acetonitrile +0.1% Trifluoroacetic acid) into each autovial inserts with samples ready to be injected and analysed on mass spectrometer.

2.6 Proteomic data of A1 and A2 HFAs using liquid chromatography – mass spectrometry (LC/MS/ MS) technique

The process of LC/MS/MS technique was optimised with multiple steps using different procedures. The first step of optimisation was the technique used to extract the protein from A1 and A2 HFAs. The following steps were taken to improve the protein extraction from the astrocytes, after analysing the chromatograms (**Figures 4** and **5**), obtained from the proteomics laboratory, UTS.

i. Clumping of the peptide-peaks was observed due to the effect of surfactant-like substances, as shown in **Figure 4**. Therefore, 1% sodium deoxycholate (SDC) was used as lysis buffer, instead of reconstitution buffer.

ii. The number of astrocytes for extraction of the proteins was increased from 1x 10^6 to 1x 10^7.

The optimisation of protein-elution in relation to the retention time was observed in the following **Figures 4** and **5**.

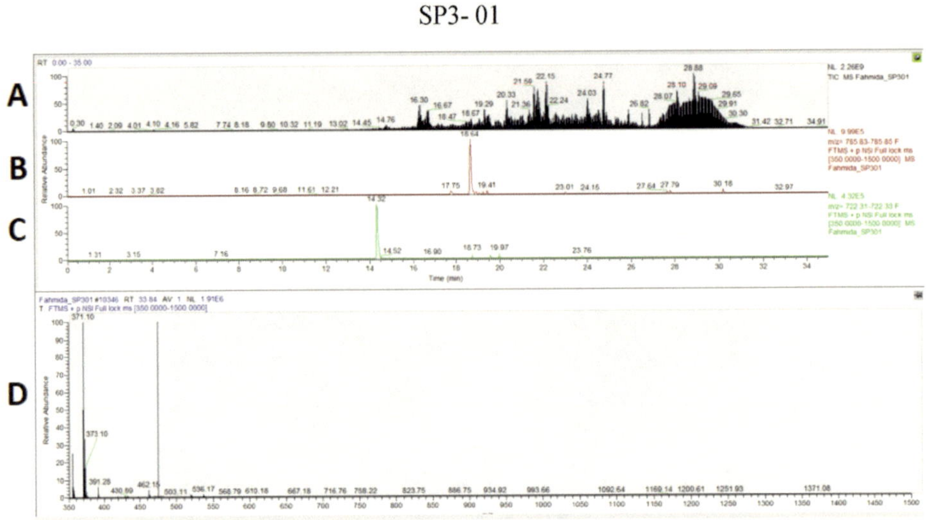

Figure 4.
There were 4 panels in this first set of chromatograms obtained after LC–MS analysis of A1 and A2 astrocytes. Panel A, shows the retention time against frequency of the peptide-peaks, which were clumped together due to surfactant-like substances. Panel B, shows the relative abundance of peptide-spikes against time. Panel C, displays the expanded version of panel B. panel D, indicates mass-to-charge ratio (m/z), used for the standardisation of the other graphs.

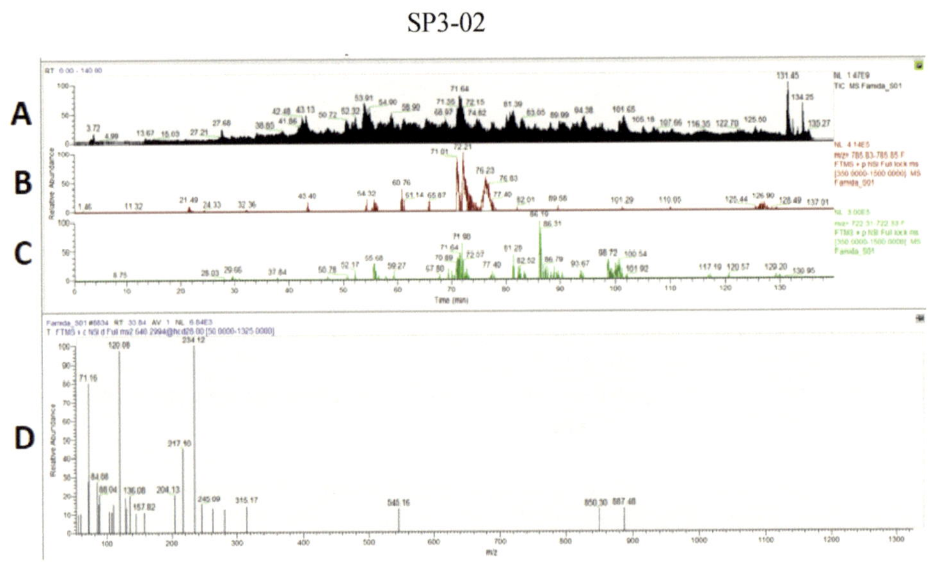

Figure 5.
Panels A, B and C with the same descriptions as for ***Figure 4****, except displaying greater retention of protein spikes, because of the optimization of protein extraction method. Panel D, indicates mass-to-charge ratio (m/z), used for the standardisation of other graphs.*

2.6.1 PEAKS software

PEAKS (Bioinformatics Solutions Inc.) software was used for the data analysis of the proteins detected from LC–MS–MS, by integrating the area under the curve (**Figure 6**) and statistical comparison of the initial data (**Figure 7**), using Student's dependent t-test as mentioned below. Triplicates of each set of A1 and A2 samples (n = 9) were carefully analysed using either one of the two filters:

i. Peaks Peptide Score (-10lgP) threshold, only peptides with a score above this threshold were used to quantify the identified proteins, separating low scoring proteins as unique peptide count.

ii. False discovery rate (FDR) threshold (1%).

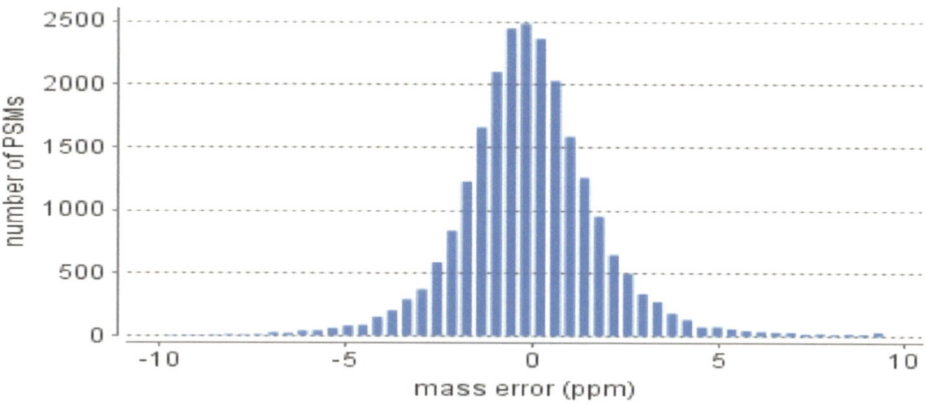

Figure 6.
Experimental error chart, indicating the minimal error (1%) in peptides-spectrum matches (PSMS) against part per million (ppm), using PEAKS software.

2.7 Statistical analysis

All values in this text are expressed as mean ± standard error mean (SEM). Differences between two groups were determined using Student's paired dependent t-test and were considered significant at $P < 0.05$. All statistical analyses were conducted using Microsoft Excel and GraphPad Prism version 8.0.1, GraphPad Software. Light microscopic micrographs were obtained on a Leica Microsystems, LAS version 4.4.0.

3. Results

3.1 A1 (reactive) and A2 (normal) HFAs in culture

The results show that the HFAs increase in size and number with increasing time in tissue culture. Distinct differences were observed in the morphology of the cells present in A1 and A2 flasks, for example A1 astrocytes have more fibroblast-like structures as compared to A2 astrocytes, which look more like neurons (**Figure 7**). A noticeable difference was observed in the thickness of the processes between the two types of the HFAs. This thickening was due to the accumulation of filament proteins such as glial fibrillary acidic proteins (GFAP) as shown in **Figure 8**. Whereas **Figure 9** shows the astrocytes present in the spontaneously hypertensive rat (SHR) and normal Wistar Kyoto (WKY) rat brain slices. **Figure 9**, shows the similarities between A1 HFAs and the SHR astrocytes, implying that the A1 HFAs mimic the hypertensive condition.

3.2 Identification of proteins in A1 and A2 HFAs, using LC/MS/MS technique

The present study was focused on specific proteins, such as glial fibrillary acidic protein (GFAP), calpain, calpastatin, cathepsin and mitogen activated protein kinase (MAPK).

3.2.1 GFAP

The GFAP protein is an indicator of HFAs reactivity, therefore, it was important to carefully measure the area under the curve to determine its degree of intensity. For comparative analysis, the peaks closest to the center of the peptides-spectrum matches (PSMS) were selected, against part per million (ppm) as shown above in **Figure 6**.

The elevated level of GFAP in A1 cells (4.39 ± 0.4) was observed, compared to A2 cells (3.02 ± 0.3), as shown in **Figure 9**. A significantly higher level of GFAP ($P < 0.05$) in A1 compared to A2, resulted in the thickening of processes (indicated by arrows in **Figures 7, 9** and **10** above).

Protein molecules such as calpain, calpastatin, cathepsin and mitogen activated protein kinase (MAPK) are known to be activated by intracellular calcium, leading to their relevant cell signalling cycles and in turn gene [24] modifications.

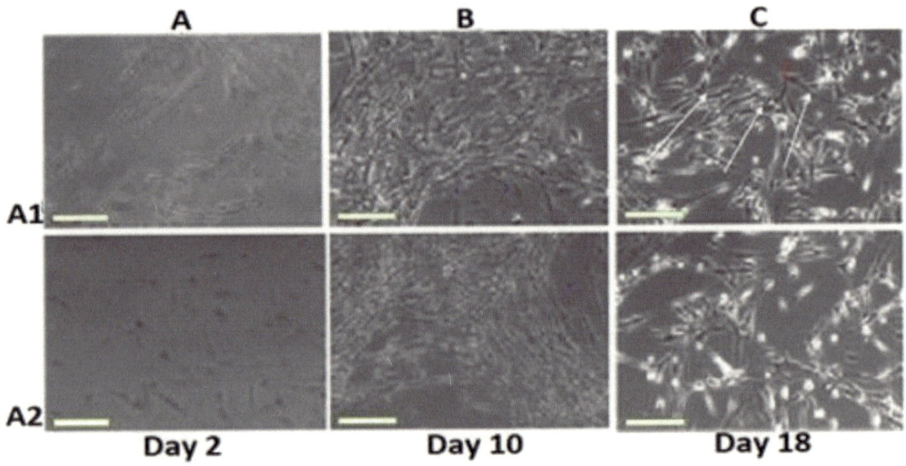

Figure 7.
Shows normal HFAs in two flasks, labelled A1 and A2. Panel A and B display 2 and 10 days old normal HFAs, respectively. Panel C, represents 18 days old normal (A2) and ATP-treated reactive (A1) astrocytes, arrows indicating the thickening of the internal filaments. Scale bar, 100 μm.

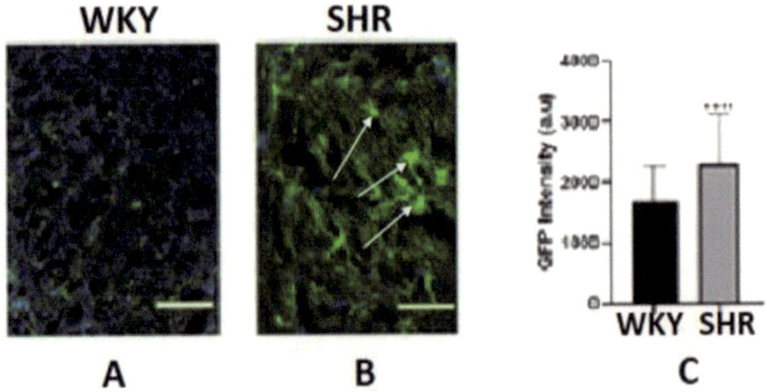

Figure 8.
*Panel A, shows the normal astrocytes in WKY rat brain slices. Panel B, displays the reactive astrocytes in SHR brain slices. Arrows indicate the upregulation of intermediate filament proteins in SHR astrocytes. Panel C, shows the difference between the intensity of GFAP in the astrocytes of normal and hypertensive rats. ***P < 0.001, scale bar, 20 μm.*

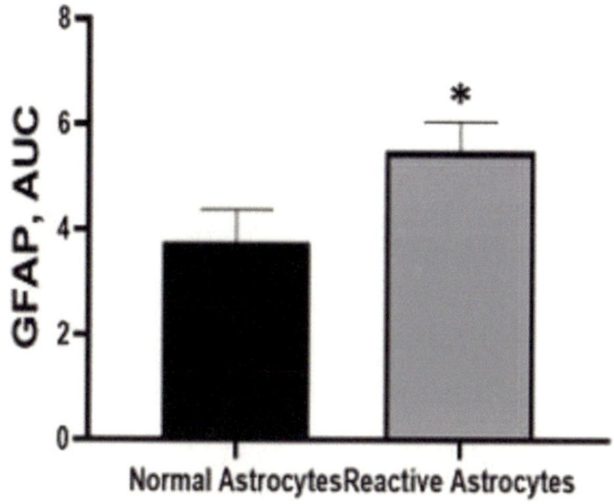

Figure 9.
*Level of glial fibrillary acidic protein (GFAP) in normal and reactive astrocytes. Area under the curve (AUC) values are shown as means ± SEM, *P < 0.05, (n = 9).*

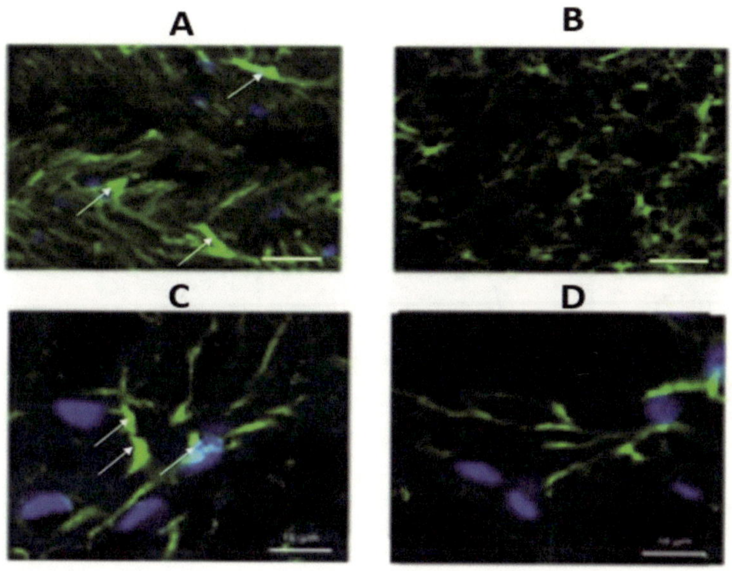

Figure 10.
Panel A (reactive) and panel B (normal) show HFAs after 10 days in culture. Panel C (reactive) and panel D (normal) is after 18 days in culture. Arrows indicate the difference in the thickening of internal filaments, i.e., on day 10 it was 2.5 μm as compared to 4.5 μm on day 18. Scale bar, 10 μm.

3.2.2 Calpain

The results show a significantly higher level (**Figure 11**) of Calpain in A1 (6.37 ± 0.6) compared to A2 (3.40 ± 0.2), indicating a greater level of calpain leading to cytotoxicity in reactive HFAs.

3.2.3 Calpastatin

The results indicate that reactive astrocytes A1 (5.52 ± 0.5) have significantly higher levels of calpastatin as compared to the normal A2 (4.09 ± 0.6) astrocytes

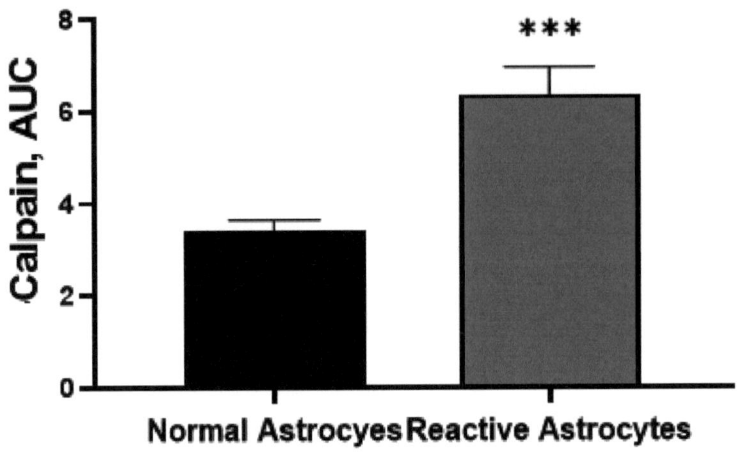

Figure 11.
Level of calpain in normal and reactive astrocytes. Area under the curve (AUC) values are shown as means ± SEM, $^{***}P < 0.001$, ($n = 9$).

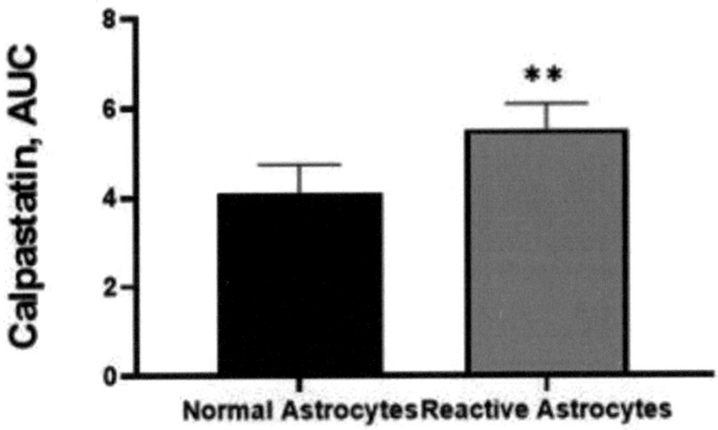

Figure 12.
Level of calpastatin in normal and reactive astrocytes. Area under the curve (AUC) values are shown as means ± SEM, $^{**}P < 0.01$, ($n = 9$).

(**Figure 12**). Calpastatin is the most specific endogenous calpain-inhibitor, which binds to the active sites of calpain to prevent its activation [25].

3.2.4 Cathepsin

A highly significant increase was observed in the levels of cathepsin in A1 (6.31 ± 0.7) compared to A2 (3.86 ± 0.4) cells (**Figure 13**). Both types of HFAs display human specific cathepsin.

3.2.5 Mitogen activated protein kinase (MAPK)

SP3 Proteomic results show a significantly higher level of MAPK enzyme in A1 (6.31 ± 0.7) compared to A2 (3.72 ± 0.4) HFAs (**Figure 14**). This kinase is an essential component of the cell signalling pathway, responsible for the communication between a receptor on the cell-surface to the DNA inside the nucleus [6, 26].

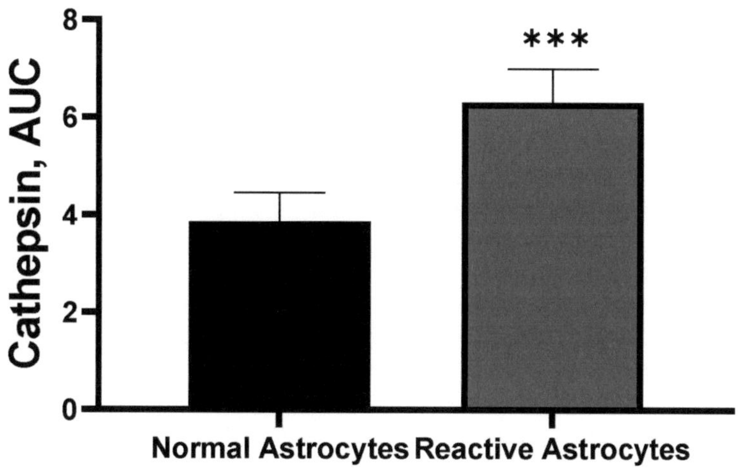

Figure 13.
*Level of cathepsin in normal and reactive astrocytes. Area under the curve (AUC) values are shown as means ± SEM, ***P < 0.001, (n = 9).*

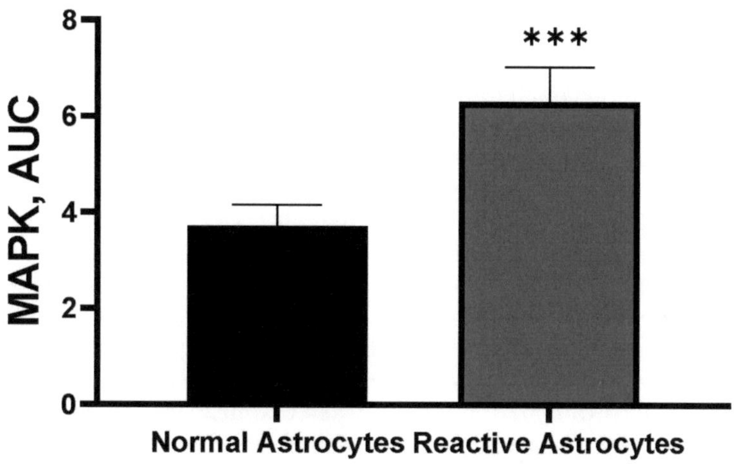

Figure 14.
*Level of MAPK in normal and reactive astrocytes. Area under the curve (AUC) values are shown as means ± SEM, ***P < 0.001, (n = 9).*

All the above-mentioned proteins were identified in both A1 and A2 HFAs, using immunocytochemistry and proteomics (SP3) protocols. Higher levels of calcium-gated proteins in A1 are potential biomarkers of HTN.

4. Discussion

This study demonstrated for the first time that reactive astrocytes in HFAs mimic hypertensive conditions. The present research indicates that the reactive (A1) HFAs have a similar protein profile to astrocytes of hypertensive rats. ATP was used in these experiments to mimic endogenous conditions, as it is released from the damaged cells following tissue injury [27, 28].

The proteins selected as predictive biomarkers for HTN, are either responsible for the reactivity of the astrocytes, or produced because of the interaction between orexin and NMDA receptors. Most of the proteins selected as predictive biomarkers were calcium binding proteases. The results from A1 and A2 HFAs show that calcium dependent proteins, such as GFAP, calpain, calpastatin, cathepsin and mitogen activated protein kinase (MAPK) were present at higher concentrations in reactive (A1) than in non-reactive (A2) astrocytes. These results suggest that the above-mentioned proteins may be targeted as therapeutic agents for the prevention of HTN.

Immunocytochemistry, in conjunction with the confocal microscopy were used to assess and visualise the proteins, such as internal filament proteins in HFAs. Mass spectrometry (LC/MS/MS) was used to validate the results, obtained by using immunocytochemistry. The SP3 proteomic protocol was employed to identify and estimate the protein-peptide molecules.

Reactive astrocytes (A1) were prepared from the normal (A2) HFAs, by using ATP (**Figure** 7), as described earlier in primary cell lines of neonatal rats [29]. In contrast, a recent research [23] reported that ATP was not sufficient to induce a complete reactive phenotype of astrocytes, in vitro. This discrepancy between the two results may be because HFAs were used in the current study, whereas Adzic et al. [23] used mature astrocytes obtained from the rats. This concept was further verified by the micrographs from light microscopy (**Figure** 7) and confocal microscopy (**Figures 8** and **10**) which showed a gradual increase in the confluency as well as in the thickness of the processes of the HFAs with increasing time in culture. Hence, our experiments confirmed that the reactivity profile of the astrocytes changed as they developed. This proliferation of the filaments is due to an increase in the accumulation of GFAP with time [30]. In A1 astrocytes, the thickness of the processes on day 10 was 2.5 μm as compared to 4.5 μm on day 18 (**Figures** 7 and **10**), indicating that there was a periodic increase in the concentration of GFAP in the reactive astrocytes. This significant thickening of astrocytic processes in A1 HFAs (**Figure 8**), was like that observed in the astrocytes of hypertensive rats (**Figure 9**). SP3 proteomic result for GFAP (**Figure 9**) was similar to the microscopic result, representing a significant increase of GFAP in reactive as compared to normal astrocytes. These results of the protein analysis were similar to the study by Hol et al. [31], indicating that an upregulation of GFAP in reactive astrocytes, coincides with neurodegenerative diseases such as HTN. Similarly, a previous study [32] has established that increased levels of GFAP was directly related to hypertension, in SHRs. Hence, based on the literature and our study results, it can be concluded that the reactivity in astrocytes could be a direct indicator of hypertension.

The results of our study showed that A1 has both increased levels of GFAP (**Figure 9**) and calpain (**Figure 11**), indicating the increased reactivity in A1 astrocytes may be due to the higher level of calpain, as shown in the previous studies [33, 34], resulting in atherosclerosis and hence HTN [35]. Other studies [36] indicate that calpain can also act on the extracellular substrates, such as collagen-fibronectin, to modulate the cell activity. As calpain can act both intra and extracellularly, it could be a therapeutic target for the prevention of cardiovascular diseases, including hypertension.

In contrast, calpastatin, an endogenous inhibitor of calpain, attenuates its cytotoxicity, thus increasing the levels of calpastatin may be beneficial in the regulation of HTN [25]. The current results show lower expression of calpastatin (**Figure 12**) as compared to calpain (**Figure 11**) in A1 astrocytes, indicating a decreased level of calpastatin in reactive HFAs. These results also suggest that an additional amount of exogenous calpastatin may decrease the toxic activity of calpain, by blocking the active sites of calpain [37]. This remarkable role of calpastatin makes it a promising therapeutic agent for managing blood pressure.

Likewise, another endogenous protease cathepsin, which is involved in inflammatory disorders, was found to be significantly higher (**Figure 13**) in reactive

(A1) astrocytes. Evidence suggests that cathepsin plays an important role in HTN through vascular modulation [38], leading to atherosclerosis. It has also been reported that cathepsin regulates the phosphorylation of mitogen activated protein kinase- kinase (MEK) in Angiotensin II-dependent hypertension [39]. Angiotensin II induces g-protein regulated MAPK cell signalling cycle [40], therefore MAPK was measured in both A1 and A2 astrocytes (**Figure 14**), and it was found to be significantly higher in A1 as compared to A2. This augmented level of MAPK in A1 HFAs could be responsible for the thickening of their intermediate filaments (**Figure 10**), as indicated previously in another study [41] that MAPK upregulates cell proliferation.

In a recent study, the phosphorylation of MAPK in the arteries of hypertensive patients as well as in a mice model [39] was observed. Moreover, the proliferation of arterial smooth muscle cells in both, human and mice models were detected. Thus, a vascular modulatory role of MAPK, which is related to HTN, was confirmed.

Our initial findings based on immunocytochemistry and SP3 proteomic results indicate that identifying molecules such as calpain, calpastatin, cathepsin and MAPK may be useful in reducing the reactivity in the astrocytes, which is an indicator of HTN. The present study shows that in A1, there is an increased levels of calcium gated proteins, possibly due to higher concentration of the intracellular calcium [42, 43], leading to hypertension.

Protein-based therapeutic agents have been highly successful in clinics [44]. More than hundred original and modified therapeutic proteins are used up till now. There are at least five ways of utilising proteins as therapeutic agents:

a. replacing damaged proteins [45]

b. augmenting the effects of weak proteins [46]

c. interfering with an intermediate molecule or its function [47]

d. providing synthetic molecules for normal function [48]

e. delivering other compounds or proteins [49]

Further experiments are needed to test these enzymes, using ELISA kits for the specific regulatory proteins, such as S100B, soluble receptor for advanced glycation end product (sRAGE) and GFAP, to identify the most promising predictive biomarker for HTN. The highly specific indicator of astrocytic reactivity, GFAP [50], was measured earlier by immunocytochemistry, proteomics, and now ELISA protocol will be used for further quantification of cytotoxicity due to the higher concentration of GFAP in the reactive astrocytes.

Similarly, the Ca^{2+}-binding protein S100B, acts both intra and extracellularly in HFAs. Inside the cell, S100B acts as a stimulator of proliferation and activation of astrocytes. Whereas, extracellularly, S100B engages sRAGE in pro-proliferative activities [51]. Though, sRAGE might not be the only S100B receptor, and S100B's ability to engage sRAGE might be regulated by its interaction with other calcium-binding proteases, such as calpain, calpastatin, cathepsin and MAPK.

5. Conclusion

This study shows for the first time that reactive astrocytes in HFAs mimic hypertensive conditions, and calcium-dependent proteins such as GFAP, calpain, calpastatin, cathepsin and MAPK could be considered as potential predictive biomarkers for HTN.

6. Future plans

More experiments are needed for the validation of the predictive biomarkers for HTN. We need to further characterise proteins (S100B, sRAGE and GFAP) in both types of HFAs, using ELISA technique.

Acknowledgements

I would like to thank my supervisors Professor Ann Simpson, Professor Kaneez Fatima Shad and Associate Professor Sara Lal for their constant support and guidance. Special thanks to Associate Professor Louis Cole and Dr. Christian Evenhuis for providing confocal microscopy and OMERO software facilities, to identify peripheral markers of HTN in HFAs, WKY and SHR. I would also like to thank Dr. Matt Padula for providing facilities to perform SP3 experiments and LC/MS/MS detection of HFA-proteins. I would like to acknowledge "Australian Government Research Training Program Scholarship".

Author details

Fahmida Abdi, Ann M. Simpson, Sara Lal and Kaneez Fatima Shad*
Life Sciences, Faculty of Science, University of Technology Sydney, Sydney, Australia

*Address all correspondence to: ftmshad@gmail.com

References

[1] Antonakoudis, G., et al., *Blood pressure control and cardiovascular risk reduction.* Hippokratia, 2007. **11**(3): p. 114-119.

[2] Shere, A., O. Eletta, and H. Goyal, *Circulating blood biomarkers in essential hypertension: a literature review.* Journal of Laboratory and Precision Medicine, 2017. **2**(12).

[3] Meissner, A., *Hypertension and the brain: A risk factor for more than heart disease.* Cerebrovascular Diseases, 2016. **42**(3-4): p. 255-262.

[4] Giles, T., *Biomarkers, cardiovascular disease, and hypertension.* The Journal of Clinical Hypertension, 2013. **15**(1): p. 1-1.

[5] Li, D.-P. and H.-L. Pan, *Glutamatergic regulation of hypothalamic Presympathetic neurons in hypertension.* Current Hypertension Reports, 2017. **19**(10): p. 78.

[6] Parsons, M.P. and L.A. Raymond, *Extrasynaptic NMDA receptor involvement in central nervous system disorders.* Neuron, 2014. **82**(2): p. 279-293.

[7] Gao, M., et al., *1-Aminocyclopropanecarboxylic acid, an antagonist of N-methyl-D-aspartate receptors causes hypotensive and antioxidant effects with upregulation of heme oxygenase-1 in stroke-prone spontaneously hypertensive rats.* Hypertension Research, 2007. **30**(3): p. 249-257.

[8] Kirchhoff, F., *Analysis of functional NMDA receptors in astrocytes.* Methods Mol Biol, 2017. **1677**: p. 241-251.

[9] Kommers, T., et al., *Phosphorylation of glial fibrillary acidic protein is stimulated by glutamate via NMDA receptors in cortical microslices and in mixed neuronal/glial cell cultures prepared from the cerebellum.* Brain Res Dev Brain Res, 2002. **137**(2): p. 139-148.

[10] Marina, N., et al., *Astrocytes monitor cerebral perfusion and control systemic circulation to maintain brain blood flow.* Nature Communications, 2020. **11**(1): p. 131.

[11] Mishra, A., *Binaural blood flow control by astrocytes: Listening to synapses and the vasculature.* The Journal of physiology, 2017. **595**(6): p. 1885-1902.

[12] Marina, N., et al., *Control of sympathetic vasomotor tone by catecholaminergic C1 neurones of the rostral ventrolateral medulla oblongata.* Cardiovasc Res, 2011. **91**(4): p. 703-710.

[13] Verkhratsky, A., R.K. Orkand, and H. Kettenmann, *Glial calcium: Homeostasis and signaling function.* Physiological Reviews, 1998. **78**(1): p. 99-141.

[14] Araque, A., et al., *Tripartite synapses: Glia, the unacknowledged partner.* Trends Neurosci, 1999. **22**(5): p. 208-215.

[15] Andres, A.L., et al., *NMDA receptor activation and Calpain contribute to disruption of dendritic spines by the stress neuropeptide CRH.* The Journal of Neuroscience, 2013. **33**(43): p. 16945.

[16] Hardingham, G.E., Y. Fukunaga, and H. Bading, *Extrasynaptic NMDARs oppose synaptic NMDARs by triggering CREB shut-off and cell death pathways.* Nature Neuroscience, 2002. **5**(5): p. 405-414.

[17] Hardingham, G.E. and H. Bading, *Synaptic versus extrasynaptic NMDA receptor signalling: Implications for neurodegenerative disorders.* Nature Reviews Neuroscience, 2010. **11**(10): p. 682.

[18] Wang, Y., et al., *Distinct roles for μ-calpain and m-calpain in synaptic NMDAR-mediated neuroprotection and extrasynaptic NMDAR-mediated neurodegeneration.* The Journal of neuroscience : the official journal of the Society for Neuroscience, 2013. **33**(48): p. 18880-18892.

[19] Karpova, A., et al., *Encoding and transducing the synaptic or extrasynaptic origin of NMDA receptor signals to the nucleus.* Cell, 2013. **152**(5): p. 1119-1133.

[20] Krapivinsky, G., et al., *The NMDA receptor is coupled to the ERK pathway by a direct interaction between NR2B and RasGRF1.* Neuron, 2003. **40**(4): p. 775-784.

[21] Bading, H., *Nuclear calcium signalling in the regulation of brain function.* Nat Rev Neurosci, 2013. **14**(9): p. 593-608.

[22] Pekny, M. and M. Pekna, *Astrocyte reactivity and reactive astrogliosis: Costs and benefits.* Physiol Rev, 2014. **94**(4): p. 1077-1098.

[23] Hughes, C.S., et al., *Single-pot, solid-phase-enhanced sample preparation for proteomics experiments.* Nature Protocols, 2019. **14**(1): p. 68-85.

[24] Adzic, M., et al., *Extracellular ATP induces graded reactive response of astrocytes and strengthens their antioxidative defense in vitro.* Journal of Neuroscience Research, 2017. **95**(4): p. 1053-1066.

[25] Zhang, S.J., et al., *Nuclear calcium signaling controls expression of a large gene pool: Identification of a gene program for acquired neuroprotection induced by synaptic activity.* PLoS Genet, 2009. **5**(8): p. e1000604.

[26] Wan, F., et al., *Extracellular Calpain/Calpastatin balance is involved in the progression of pulmonary hypertension.* Am J Respir Cell Mol Biol, 2016. **55**(3): p. 337-351.

[27] Karpova, A., et al., *Encoding and Transducing the Synaptic or Extrasynaptic Origin of NMDA Receptor Signals to the Nucleus*. Vol. 152. 2013. 1119-33.

[28] Khakh, B.S. and R.A. North, *Neuromodulation by extracellular ATP and P2X receptors in the CNS.* Neuron, 2012. **76**(1): p. 51-69.

[29] Boué-Grabot, E. and Y. Pankratov, *Modulation of central synapses by astrocyte-released ATP and postsynaptic P2X receptors.* Neural plasticity, 2017. **2017**: p. 9454275-9454275.

[30] Neary, J.T., et al., *Extracellular ATP induces stellation and increases glial fibrillary acidic protein content and DNA synthesis in primary astrocyte cultures.* Acta Neuropathologica, 1994. **87**(1): p. 8-13.

[31] Cho, W. and A. Messing, *Properties of astrocytes cultured from GFAP over-expressing and GFAP mutant mice.* Experimental cell research, 2009. **315**(7): p. 1260-1272.

[32] Hol, E.M. and M. Pekny, *Glial fibrillary acidic protein (GFAP) and the astrocyte intermediate filament system in diseases of the central nervous system.* Current Opinion in Cell Biology, 2015. **32**: p. 121-130.

[33] Tomassoni, D., et al., *Increased expression of glial fibrillary acidic protein in the brain of spontaneously hypertensive rats.* Clinical and Experimental Hypertension, 2004. **26**(4): p. 335-350.

[34] Lee, Y., et al., *Rapid increase in immunoreactivity to GFAP in astrocytes in vitro induced by acidic pH is mediated by calcium influx and calpin I.* Brain research, 2000. **864**: p. 220-229.

[35] Kim, J.H., et al., *Reactive protoplasmic and fibrous astrocytes contain high levels of calpain-cleaved alpha 2 spectrin.* Exp Mol Pathol, 2016. **100**(1): p. 1-7.

[36] Miyazaki, T., et al., *M-Calpain induction in vascular endothelial cells on human and mouse atheromas and its roles in VE-cadherin disorganization and atherosclerosis.* Circulation, 2011. **124**(23): p. 2522-2532.

[37] Letavernier, B., et al., *Calpains contribute to vascular repair in rapidly progressive form of glomerulonephritis: Potential role of their externalization.* Arterioscler Thromb Vasc Biol, 2012. **32**(2): p. 335-342.

[38] Wendt, A., V. Thompson, and D. Goll, *Interaction of calpastatin with calpain: A review.* Biological chemistry, 2004. **385**: p. 465-472.

[39] Cheng Xian, W., et al., *Cysteine protease Cathepsins in atherosclerosis-based vascular disease and its complications.* Hypertension, 2011. **58**(6): p. 978-986.

[40] Lu, Y., et al., *Angiotensin II-induced vascular remodeling and hypertension involves cathepsin L/V- MEK/ERK mediated mechanism.* International Journal of Cardiology, 2020. **298**: p. 98-106.

[41] Goldsmith, Z.G. and D.N. Dhanasekaran, *G protein regulation of MAPK networks.* Oncogene, 2007. **26**(22): p. 3122-3142.

[42] Schevzov, G., et al., *Regulation of cell proliferation by ERK and signal-dependent nuclear translocation of ERK is dependent on Tm5NM1-containing actin filaments.* Molecular biology of the cell, 2015. **26**(13): p. 2475-2490.

[43] Hazari, M.A., et al., *Serum calcium level in hypertension.* North American Journal of Medical Sciences, 2012. **4**(11): p. 569-572.

[44] Villa-Etchegoyen, C., et al., *Mechanisms involved in the relationship between low calcium intake and high blood pressure.* Nutrients, 2019. **11**(5): p. 1112.

[45] Dimitrov, D.S., *Therapeutic proteins.* Methods in molecular biology (Clifton, N.J.), 2012. **899**: p. 1-26.

[46] Gorzelany, J.A. and M.P. de Souza, *Protein Replacement Therapies for Rare Diseases: A Breeze for Regulatory Approval?* Science Translational Medicine, 2013. **5**(178): p. 178fs10.

[47] Carter, P.J., *Introduction to current and future protein therapeutics: A protein engineering perspective.* Experimental Cell Research, 2011. **317**(9): p. 1261-1269.

[48] Baker, M.P. and T.D. Jones, *Identification and removal of immunogenicity in therapeutic proteins.* Current opinion in drug discovery & development, 2007. **10**(2): p. 219-227.

[49] Goeddel, D.V., et al., *Expression in Escherichia coli of chemically synthesized genes for human insulin.* Proceedings of the National Academy of Sciences, 1979. **76**(1): p. 106.

[50] Emi Aikawa, N., et al., *Immunogenicity of anti-TNF-α agents in autoimmune diseases.* Clinical Reviews in Allergy & Immunology, 2010. **38**(2): p. 82-89.

[51] Middeldorp, J. and E.M. Hol, *GFAP in health and disease.* Progress in Neurobiology, 2011. **93**(3): p. 421-443.

Chapter 5

Detection of Benzo[a]Pyrene Diol Epoxide-DNA Adducts in White Blood Cells of Asphalt Plant Workers in Syria

Razan Zohairee

Abstract

Benzo[a]pyrene (B[a]P) is a major polycyclic aromatic hydrocarbon (PAH), it can bind the DNA to produce DNA-adducts, which has major carcinogenic potential. Enzyme-linked immunosorbet assay (ELISA) is the method used to detect these DNA adducts of B[a] P diolepoxide (BPDE) within the living cells. The aim of this study is to evaluate exposure to bitumen fumes, and to B[a]P in asphalt plant workers by measuring the BPDE-DNA adducts in their peripheral white blood cells (WBC), which are considered biological markers for exposure risk assessment. In this study, Hemostatic blood (CBC, AST) were measured, and the levels of BPDE-DNA adducts were measured in DNA samples of WBC obtained from asphalt plant workers in Syria and compared to those measured from a control group. The measurement was performed using BPDE-DNA Adducts ELISA.kit. The sample size was determined to be 50 with 25 asphalt plant workers and 25 healthy volunteers with no occupational exposure to PAHs. The results showed some diseases associated with exposure to asphalt fumes among the workers in the study group and a statistically significant difference in the values of (CBC; WBC, leukocytes, HCT, MCHC and AST) between the study group and the control group. BPDE-DNA adducts were detected in WBC of 11 asphalt plant workers with concentrations ranging between 0 and 2.75 ng/ml and only one individual in the control group with concentration of 0.75 ng/ml. These results indicate significant positive relationship between exposure to the bitumen fumes and formation of BPDE-DNA adducts. BPDE-DNA adducts is potential biomarker for PAHs exposure and likely helpful indicator of PAH-induced DNA damage and possibly carcinogenesis.

Keywords: benzo[a]pyrene, CBC, AST, BPDE-DNA adducts, asphalt workers

1. Introduction

Workers are exposed during the asphalt industry to bitumen fumes that contain poly aromatic hydrocarbons (PAHs) [1], organic vapors, silica dust, diesel exhaust, asbestos fibers and coal tar [2].

Exposure to bitumen fumes causes headache, fatigue, lack of appetite, eye, skin and bronchitis irritation, coughing, bronchitis, asthma [3], genotoxic effects [4], damaging DNA [5], cancer lung [6], increases the risk of developing bladder cancer [7].

PAHs are known environmental pollutants with harmful effect on human health. Benzo [*a*] pyrene (B[a]P) is a lead compound in this group and one of the most studied carcinogenic PAHs [8]. B[a]P is formed during incomplete combustion of organic materials and pyrolysis of inorganic compounds [9] B[a]P is also found in cigarette smoke [10], cooked food [11], and various combustion gases such as vehicle exhaust [12]. It is also generated from some industrial operations such as those of cooking ovens, heavy oil plants [13] and asphalt plants [14]. Given its harmful effects on health, B[a]P was classified by the International Agency for Research on Cancer (2012) among the highly genotoxic compounds and categorized to "group 1 carcinogenic to humans" [15].

Asphalt workers are exposed to B[a]P through two major ways: though inhalation of emanating fumes from the chimney for those working in the mixing plants, or through inhalation of the vaporous gas while paving asphalt at roadside workshops [16]. Notably, when asphalt is being prepared for road paving, the mixture reaches very high temperature ranging between 130° and 145°, which results emission of large of amount of vaporous fumes, that workers inadvertently inhale. To a lesser extent, some workers are exposed to B[a]P via touching B[a]P-containing compounds [16]. This particularly occurs among mechanics that are often exposed to petroleum compounds. According to the occupational safety and health administration, the allowed concentration of B[a]P in environment when workers perform their duty should not exceed 0.2 mg/ and exposure should not exceed more than 8 hours a day [17].

Although B[a]P is metabolized in all human tissues, the hepatic catabolism of B[a]P through cytochrome P450 (specifically cytochrome P450 1A1) results in production of several metabolites such as epoxides, dihydrodiols, phenols and quinones. First, B[a]P undergo phase I metabolism into epoxides. Subsequently, these epoxides undergo further metabolism though one of three ways: through spontaneous rearrangement to phenols, through hydration to transdihydrodiols by epoxide hydrolase or though phase II detoxification. Phase II dextoxification occurs via binding to glutathione, which can occur spontaneously or though catabolism by glutathione-S-transferases [18]. Given this extensive hepatic metabolism, the amount of B[a]P that reaches the systemic circulation is minimal. Therefore, gastrointestinal route appears to be a less important route of exposure to B[a]P.

B[a]P appears a carcinogenic compound which is metabolized by cytochrome 450 1A1 to form DNA-adducts, that play key roles in its carcinogenesis [9].

The mechanism of carcinogenesis of B[a]P is dependent on a 3-step enzymatic metabolism discussed above (**Figure 1**) [18]. These steps eventually form the final mutagen BPDE [19], which is a very reactive and binds covalently to lipids, protein and DNA to form BPDE adducts [20, 21]. These BPDE-DNA adducts can result in permanent mutations if not promptly repaired, which can lead to development of cancer.

In this study, we examine the relationship between being an asphalt worker and detection of BPDE-DNA adducts in the white blood cells (WBC), which can be a useful surrogate exposure risk.

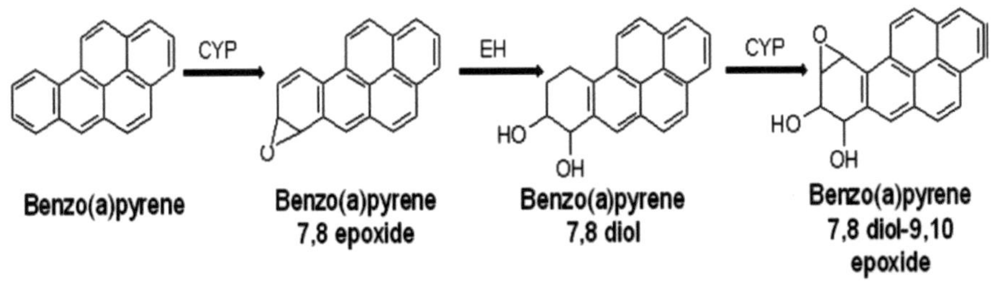

Figure 1.
B[a]P converts to the final carcinogen, BPDE.

2. Subjects and methods

This was a cross-sectional study that aimed to evaluate the exposure to asphalt fumes (i.e., by measure the hemostatic blood CBC, and Aspartate Amino Transeferase AST), and to evaluate the exposure to B[a]P through asphalt fumes among asphalt workers by determining the levels of BPDE-DNA adducts in their WBC. It was conducted on 25 male workers exposed to asphalt fumes and 25 healthy male individuals without such exposure.

2.1 Comparison groups

Group 1 (study group): This group consisted of 25 male individuals who worked at an asphalt plant in Syria. Among those, 12 subjects worked in the mixing plant (9 were employed to monitor the equipment operation and 3 mechanics to maintain the equipment). The remainder 13 subjects were paving workers at roadside workshops (5 subjects to operate the equipment, 4 mechanics and 4 manual construction workers). These subjects were exposed to B[a]P vapors and aerosols of bitumen for many years ranging from 3 to 31 years. On average, all subjects worked 8 hours a day, 4 days a week. Their mean age was 46 years (range: 33–58 years).

The subjects within the study group were carefully selected to avoid external bias. All subjects did not drink alcohol. They all had somewhat comparable dietary habits, that consisted predominantly of vegetables. All subjects did not work outside their primary site of employment. Subjects with significant non-occupational exposure to PAHs (e.g. use burning woods for heat) or those who work a second job were excluded.

Group 2 (control group): The control group consisted of 25 healthy volunteers. These individuals were predominantly university administrators and hospital workers. They did not work outside their primary site of employment. There all had no significant occupational or personal exposure to PAHs (e.g. did not use burning woods for heat). They all took no prescription medications and lived remotely from bitumen fumes. The subjects in the control group were selected to match the subjects in exposure group in gender, age, smoking status and dietary habits.

Subgroups: Subjects in the study group were categorized into subgroups based on age, and smoking status. Based on age, they were categorized by age: ˂45 years (8 workers, ranging from 33–44 years) and ≥ 45 years (17 workers, ranging from 46–58 years). Based on smoking status, they were categorized to smokers (22 subjects) and non-smokers (3 subjects).

Study protocol: All subjects signed written consent prior to enrollment in the study. The study protocol was approved by the Ethical Committee of Damascus University prior to the start of the study.

The study was begun with a questionnaire, that was administered to all subjects to collect general personal information such as age, weight, health status, smoking status, dietary and lifestyle habits and other demographic data.

After enrollment of qualified subjects, blood samples were collected via cubital venipuncture and 5 mL of blood were collected from each subject and placed in sterile tubes. For each subject, one of these tubes was placed immediately in ice bags at 4° and transported immediately to the hospital to be stored in the freezer at −80° until the staff were ready to analyze it. The remainder of the specimen was used immediately to perform various laboratory biomarkers such as complete blood count CBC; WBC, red blood cells RBC, hematocrit (HCT), hemoglobin (HGB) and mean corpuscular hemoglobin concentration (MCHC) using (Sysmex Cooperation, Japan) and (AST) (reference range 0–35 unit/L) using (Biosystem S.A. Spain).

All blood samples were collected in April, 2016.

2.2 DNA isolation

Genomic DNA was isolated from peripheral WBC by using DNA Kit (Thermo Scientific Gene JET Genomic DNA Purification Kit) (Qiagen, USA). DNA concentrations were measured using Gene Nano Drop (Biochrom, England), which occured in the biology laboratory at Alassad University Hospital in Damascus, Syria. Repeated thawing and freezing of the samples were avoided.

2.3 Determination of BPDE-DNA adduct levels

After DNA was isolated from WBC, the samples were subsequently diluted to a concentration of 2 μg of DNA in 1 mL. Phosphate-buffered Saline with PH 7.2 (1X) (Gibco by life technologies) was used for dilution and washing. BPDE-DNA adduct levels (ng/ml) were measured according to the standard method provided by OxiSelect BPDE-DNA Adduct ELISA Kit (Cell Biolabs, Inc., San Diego, USA). This ELISA kit is an immunoassay enzyme developed for rapid detection of BPDE-DNA adducts. The quantity of BPDE adduct in DNA samples is determined by relative comparison of a known BPDE-DNA standard curve. The kit provides sufficient reagents to perform up to 96 assays, including standard curve and unknown DNA samples. The results were expressed as nanograms of BPDE-DNA adducts per microgram of DNA. The analyses were applied twice.

Apparatus: ELISA (Tecan, Switzerland) in the Blood Bank of Damascus University.

Statistical analysis: The statistical analysis of this study was performed using SPSS software version 13.0. P value of <0.05 was considered to be statistically significant. Mann–Whitney U test was used to compare continuous data with non-normal distribution, $\chi 2$ test was used to compare categorical variables between the two groups, and spearman coefficient and independent T- student tests were used.

3. Results

3.1 Medical conditions and basic laboratory data

Review of the subject's basic data revealed that some study group suffers from a health condition, particularly respiratory diseases (chest pains, shortness of breath, asthma). This was particularly true among mechanics, one of the workers has olfactory deficits. In total, 44% of asphalt plant workers had at least one medical condition.

3.1.1 Hemostatic assays

Laboratory studies showed significantly higher the number of WBC with T-student test among the exposed group compared to the control group (8.72 ± 1.92 and P = 0.015), particularly lymphocytes (32.09 ± 6.99 and P = 0.003). (HCT) and (MCHC) were significantly higher among the study group compared to the control group (P = 0.034 and P < 0.0001, respectively). The asphalt workers have significantly higher levels of (AST) (40.59 ± 7.56 unit/l, p < 0.0001).

3.1.2 BPDE-DNA adducts levels in the study group

In the study group, BPDE-DNA adducts were detected in the WBC of 11 out of 25 individuals (44%), 2 non-smokers and 9 smokers. The concentrations of

BPDE-DNA adducts among the exposed group ranges between(0 to 2.75 ng/mL) except for one individual who had exceptionally high concentration at 16.5 ng/mL. He was a smoker.

In the control group, BPDE-DNA adducts were detected in only one individual with concentration of 0.75 ng/ml (**Table 1**). He was a smoker.

Using statistical analysis, the number of individuals with detected BPDE-DNA adducts in their WBC was higher among the study group compared to the control group ($\chi 2$ test, $p = 0.001$).

3.1.3 BPDE-DNA adducts levels according to smoking status

In the study group, there was no significant difference in the number of individuals with detected BPDE-DNA adducts between smokers and non-smokers (Mann–Whitney U test, $P = 0.43$), the results of Levels of BPDE-DNA adducts for Group Study below (**Table 1**).

3.1.4 BPDE-DNA adducts levels according to age

Among asphalt workers, the concentrations of BPDE-DNA adducts ranges from 0 to 2.75 ng/mL for subjects <45 year-old and ranges from 0 to 2.25 ng/mL for those ≥45 year-old except for one subject with unusually high concentration of 16.5 ng/mL. He was from the older age category.

3.1.5 BPDE-DNA adduct levels according to disease status

As discussed above, 44% of asphalt workers had health conditions and/or symptoms such as chest pain, dyspnea, and asthma. Among the asphalt workers, there was a statistically non-significant trend toward positive relationship between the concentrations of BPDE-DNA adducts and disease status ($\chi 2$ test, $p = 0.08$). Nonetheless, one asphalt worker suffered from olfactory deficits (his BPDE-DNA adducts concentration was 2.25 ng/mL) and another suffered from spinal disc disease (his BPDE-DNA adducts concentration was 16.5 ng/mL).

3.1.6 BPDE-DNA adducts levels according various laboratory biomarkers

The relationship between BPDE-DNA adducts levels and various laboratory biomarkers (CBC variables and AST) among asphalt workers was assessed using spearman coefficient. There was no significant correlation identified between BPDE-DNA adducts levels and any of these variables.

Group Study	**Sample Number (N)**			**Percentage (%)**		
	Sum	**BPDE-DNA adducts**	**non- BPDE-DNA adducts**	**Sum**	**BPDE-DNA adducts**	**non- BPDE-DNA adducts**
Study group	25	11	14	100	44	56
Control group	25	1	24	100	4	96

Table 1.
Levels of BPDE-DNA adducts.

4. Discussion

When the responses to the administered questionnaires were reviewed, a significantly higher number of health illnesses were identified among asphalt workers, particularly respiratory diseases. They were identified in both smokers and non-smokers and therefore, they are likely attributed to exposure to bitumen fumes, where inhalation is the primary modality through which asphalt workers were exposed to it. Prior studies such as conducted by (Gamble et al.,1999) [22] align with our observation and indicate significant positive relationship between exposure to bitumen fumes and respiratory diseases.

It was also noted that mechanics that work in asphalt plants are at higher risk to develop respiratory disease [22], which could be attributed to the impact of PAHs they were exposed.

In our study, we noted changes in some of the hematological parameters such as increase in the number of lymphocytes, which aligns with the study published by (Tompa et al) [23] where may attributed this increase to the harmful effects of PAHs.

In addition, the subjects in our study group had significantly higher HCT and MCHC levels compared to the control group, which is likely due to exposure to PAHs. These findings are in alignment with the study conducted by (Wang et al) [24] which showed significantly higher hemoglobin levels with some atypical appearance of the red blood cells, which were attributed to exposure to PAHs.

We also noted higher hemoglobin levels among smokers in both groups, which is a compensatory mechanism that occurs due to decrease in the oxygen level and increase carbon monoxide levels due to smoking [25–27].

We also noted significantly higher AST level between the study and control groups, in contrast to the findings of (Atasoy et al) [28]. The contradictory observations between these two studies could be explained by difference in the type of asphalt used, or the method of use and the extend and level of exposure to PAHs.

In our study, we noted relationship between exposure to PAHs by inhaling bitumen fumes from working closely with asphalt and detection of BPDE-DNA adducts ($P = 0.001$). On the other hand, (Pavanello et al). demonstrated significant relationship between chronic inhalation of high levels of PAHs and detection of BPDE-DNA adducts [29].

Interestingly, although smoking has somewhat similar effect on DNA as exposure to B[a]P, this study showed no correlation between smoking and the levels of BPDE-DNA adducts ($p = 0.43$). These findings are in agreement with those of (Pavanello et al) [29] and (van Schooten et al) [30]. However, other studies found significant correlation between the levels of BPDE-DNA adducts and smoking in subjects with occupational exposure PAHs where exposure to tobacco products and PAHs acted synergistically to form BPDE-DNA adducts as reported by (Rojas et al) [31]. These discrepancy between these findings could be explained by inter-individual factor variability and route of exposure to PAHs which can play a major role in formation of BPDE-DNA adducts.

Statistically, there was no significant difference in the adducts concentrations between the younger and older age groups ($\chi 2$ test, $p = 0.18$). On the other hand, age did not significantly correlate with the presence of BPDE-DNA adducts among subjects in the study group ($p = 0.18$). These results are in agreement of those reported by (McClean et al) [32]. and is likely due to the similar metabolic and excretion ability in both age groups.

Given the known carcinogenic effect of B[a]P [8], the presence of BPDE-DNA adducts when measured in the WBC using ELISA technique correlates of the risk of development of B[a]P-induced diseases such as lung cancer [33]. Therefore,

BPDE-DNA adducts can serve as a useful biomarker to assess prior exposure to PAHs and could potentially determine cancer risk [34].

Our results suggests that importance of measurement of BPDE-DNA adducts and its role as potential biomarker for exposure to PAHs. As it involves the DNA, it could be a surrogate marker to assess the risk for development of cancer [35, 36].

5. Conclusions

AST levels and some vital blood parameters can be considered warning indicators for rapid diagnosis of exposure for the asphalt workers. Our results highlight that importance of measurement of BPDE-DNA adducts and its role as potential biomarker for exposure to PAHs and being considered a biomarker for unrepaired DNA damage.

As it involves the DNA, it could be a surrogate marker to assess the risk for development of cancer.

Acknowledgements

Prof. Marouf Alkayer, Professor of Analytical Chemistry at Faculty of Pharmacy at Tishreen University in Syria.

Conflict of interest

The authors declare no conflict of interest.

Author details

Razan Zohairee
Department of Toxicology and Pharmacology, Damascus Univresity, Damascus, Syria

*Address all correspondence to: rita.joo@hotmail.com

References

[1] Xu Y, Lindh CH, Jönsson BA, Broberg K, Albin M. Occupational exposure to asphalt mixture during road paving is related to increased mitochondria DNA copy number: a cross-sectional study. Environmental Health. 2018;17(1):29.

[2] Lauby-Secretan B, Baan R, Grosse Y, El Ghissassi F, Bouvard V, Benbrahim-Tallaa L, et al. Bitumens and bitumen emissions, and some heterocyclic polycyclic aromatic hydrocarbons. Elsevier; 2011.

[3] Randem B, Ulvestad B, Burstyn I, Kongerud J. Respiratory symptoms and airflow limitation in asphalt workers. Occupational and environmental medicine. 2004;61(4):367-9.

[4] Karaman A, Pirim I. Exposure to bitumen fumes and genotoxic effects on Turkish asphalt workers. Clinical Toxicology. 2009;47(4):321-6.

[5] Bacaksiz A, Kayaalti Z, Soylemez E, Tutkun E, Soylemezoglu T. Lymphocyte DNA damage in Turkish asphalt workers detected by the comet assay. International journal of environmental health research. 2014;24(1):11-7.

[6] Burstyn I, Boffetta P, Kauppinen T, Heikkilä P, Svane O, Partanen T, et al. Estimating exposures in the asphalt industry for an international epidemiological cohort study of cancer risk. American journal of industrial medicine. 2003;43(1):3-17.

[7] Burstyn I, Kromhout H, Johansen C, Langard S, Kauppinen T, Shaham J, et al. Bladder cancer incidence and exposure to polycyclic aromatic hydrocarbons among asphalt pavers. Occupational and environmental medicine. 2007;64(8):520-6.

[8] Abdel-Shafy HI, Mansour MSM. A review on polycyclic aromatic hydrocarbons: Source, environmental impact, effect on human health and remediation. Egyptian Journal of Petroleum. 2016 2016/03/01/;25(1):107-23.

[9] Rengarajan T, Rajendran P, NandakumarN,LokeshkumarB,RajendranP, Nishigaki I. Exposure to polycyclic aromatic hydrocarbons with special focus on cancer. Asian Pacific Journal of Tropical Biomedicine. 2015 2015/03/01/;5(3):182-9.

[10] Hecht SS. Cigarette smoking and lung cancer: chemical mechanisms and approaches to prevention. The lancet oncology. 2002;3(8):461-9.

[11] Hilker DM. Diet and carcinogenesis: Carcinogens occurring naturally in food. Nutrition and cancer. 1981;2(4):217-23.

[12] Gelboin HV. Benzo [alpha] pyrene metabolism, activation and carcinogenesis: role and regulation of mixed-function oxidases and related enzymes. Physiological reviews. 1980;60(4):1107-66.

[13] Yang H-H, Lee W-J, Chen S-J, Lai S-O. PAH emission from various industrial stacks. Journal of Hazardous Materials. 1998;60(2):159-74.

[14] Butler MAS, Burr G, Dankovic DA, Lunsford RA, Miller AK, Nguyen M, et al. Health effects of occupational exposure to asphalt. 2000.

[15] Ewa B, Danuta M-Š. Polycyclic aromatic hydrocarbons and PAH-related DNA adducts. Journal of applied genetics. 2017;58(3):321-30.

[16] McClean MD, Rinehart RD, Ngo L, Eisen EA, Kelsey KT, Herrick RF. Inhalation and Dermal Exposure among Asphalt Paving Workers. The Annals of Occupational Hygiene. 2004;48(8):663-71.

[17] OSHA. Occupational Safety and Health Administration 2013 [cited 2018 1-3]. Available from: https://www.osha.gov/dsg/annotated-pels/tablez-1.html.

[18] Miller KP, Ramos KS. Impact of cellular metabolism on the biological effects of benzo [a] pyrene and related hydrocarbons. Drug metabolism reviews. 2001;33(1):1-35.

[19] Klaassen CD, Amdur MO. Casarett and Doull's toxicology: the basic science of poisons. 8 ed ed: McGraw-Hill New York; 2013.

[20] Shiizaki K, Kawanishi M, Yagi T. Modulation of benzo [a] pyrene–DNA adduct formation by CYP1 inducer and inhibitor. Genes and Environment. 2017;39(1):14.

[21] Hodek P, Koblihová J, Kizek R, Frei E, Arlt VM, Stiborová M. The relationship between DNA adduct formation by benzo [a] pyrene and expression of its activation enzyme cytochrome P450 1A1 in rat. Environmental toxicology and pharmacology. 2013;36(3):989-96.

[22] Gamble JF, Nicolich MJ, Barone NJ, Vincent WJ. Exposure-response of asphalt fumes with changes in pulmonary function and symptoms. Scandinavian journal of work, environment & health. 1999:186-206.

[23] Tompa A, Jakab MG, Biró A, Magyar B, Major J. Health, genotoxicology, and immune status of road pavers in Hungary. Journal of Occupational and Environmental Hygiene. 2007;4(S1):154-62.

[24] Wang L, Zhao Y, Liu X, Huang T, Wang Y, Gao H, et al. Cancer risk of petrochemical workers exposed to airborne PAHs in industrial Lanzhou City, China. Environmental Science and Pollution Research. 2015;22(24):19793-803.

[25] Shah B, Nepal A, Agrawal M, Sinha A. The effects of cigarette smoking on hemoglobin levels compared between smokers and non-smokers. Sunsari Technical College Journal. 2013;1(1):42-4.

[26] Aitchison R, Russell N. Smoking—a major cause of polycythaemia. Journal of the Royal Society of Medicine. 1988;81(2):89-91.

[27] Malenica M, Prnjavorac B, Bego T, Dujic T, Semiz S, Skrbo S, et al. Effect of Cigarette Smoking on Haematological Parameters in Healthy Population. Medical Archives. 2017 01/05/received1/25/accepted;71(2):132-6. PubMed PMID: PMC5511531.

[28] Atasoy N, Kanat Y. Determination of the amount of certain heavy metal ions and some specific liver enzymes and levels of testosterone hormone in the blood sera of heavy asphalt workers and rural community in Van, Turkey. Research Journal of Medical Sciences. 2011;5(2):73-9.

[29] Pavanello S, Favretto D, Brugnone F, Mastrangelo G, Pra GD, Clonfero E. HPLC/fluorescence determination of anti-BPDE–DNA adducts in mononuclear white blood cells from PAH-exposed humans. Carcinogenesis. 1999;20(3):431-5.

[30] Van Schooten F, Hillebrand M, Van Leeuwen F, Van Zandwijk N, Jansen H, Engelse Ld, et al. Polycyclic aromatic hydrocarbon—DNA adducts in white blood cells from lung cancer patients: no correlation with adduct levels in lung. Carcinogenesis. 1992;13(6):987-93.

[31] Rojas M, Alexandrov K, Auburtin G, Wastiaux-Denamur A, Mayer L, Mahieu B, et al. Anti-benzo [a] pyrene diolepoxide—DNA adduct levels in peripheral mononuclear cells from coke oven workers and the enhancing effect of smoking. Carcinogenesis. 1995;16(6):1373-6.

[32] McClean M, Wiencke J, Kelsey K, Varkonyi A, Ngo L, Eisen E, et al. DNA adducts among asphalt paving workers. The Annals of occupational hygiene. 2006;51(1):27-34.

[33] Rojas M, Cascorbi I, Alexandrov K, Kriek E, Auburtin G, Mayer L, et al. Modulation of benzo [a] pyrene diolepoxide–DNA adduct levels in human white blood cells by CYP1A1, GSTM1 and GSTT1 polymorphism. Carcinogenesis41.–. 2000;21(1):35

[34] Moorthy B, Chu C, Carlin DJ. Polycyclic aromatic hydrocarbons: from metabolism to lung cancer. Toxicological sciences : an official journal of the Society of Toxicology. 2015 May;145(1):5-15. PubMed PMID: 25911656. Pubmed Central PMCID: Pmc4408964. Epub 2015/04/26. eng.

[35] Wiencke JK. DNA adduct burden and tobacco carcinogenesis. Oncogene. 2002 10/15/online;21:7376.

[36] Li D, Firozi PF, Wang LE, Bosken CH, Spitz MR, Hong WK, et al. Sensitivity to DNA damage induced by benzo(a)pyrene diol epoxide and risk of lung cancer: a case-control analysis. Cancer research. 2001 Feb 15;61(4):1445-50. PubMed PMID: 11245449. Epub 2001/03/14. eng.

Chapter 6

The Study of Some Possible Risk Factors for Arterial Thrombosis in the Example of Georgian Patients

Marina Koridze, Marina Nagervadze, Maria Sarkhaiani, Leila Akhvlediani, Rusudan Khukhunaishvili, Ketevan Dolidze, Sophiko Tskvitinidze, Shorena Gabaidze, Irina Nakashidze, Sopio Garakanidze and Giorgi Nikolaishvili

Abstract

Cardiovascular diseases (CVD) are the most common cause of death worldwide. As arterial as venous thrombosis are major cause's morbidity and mortality. There is an exponential increase in the risk of arterial and venous thrombotic events with age, gender, smoking habits, diet type etc. The association of arterial and venous thrombosis and ABO histo-blood group is well established. Our research aim was to find a possible relationship between age, gender, smoking habit, ABO, Rh, Kell, MN blood group and arterial thrombosis in the example of the Georgian population. The study material comprised the blood samples of 100 patients with arterial thrombosis. Also, control (donor) groups were studied. The Control group included individuals without cardiovascular disease during the 2019–2020 periods. They were asymptomatic, healthy persons. The immunoserological express method with universal monoclonal antibodies ware used. 77% of the studied patients were males. The majority of patients were over the 60 years old. 35% of our studied patients are non-smoker, 39% are ex-smoker, and 26% are active smokers. A similar distribution has the ABO and Rh phenotypes in patients and donors. M+ N+ (MN) the phenotype is relatively high in the patient group to comparing to donors. Kell antigen prevalence was relatively high in studied patients. Our study has shown maleness as a higher risk factor for arterial thrombosis. The smokers have a more predicted chance for arterial thrombosis. K+ phenotype and M+ N+ characteristics are a high prevalence in patients. There is no correlation between ABO and Rh blood groups with arterial thrombosis.

Keywords: risk factor, arterial thrombosis, red blood cells, ABO group, Rh group

1. Introduction

Cardiovascular diseases (CVD) are the most common causes of death worldwide. As already well known, the CVD has several risk factors such as blood pressure, glucose, lipids, blood pressure, glucose, lipids, blood pressure glucose, lipids etc. As arterial as venous thrombosis are major cause's morbidity and mortality.

IntechOpen

There is an exponential increase in the risk of arterial and venous thrombotic events with age, gender, smoking habits, diet type, etc. [1].

The reason of arterial thrombosis (AT) is rupture of atherosclerotic plaque and formation of platelet-mediated thrombi, which can cause ischemic injuries. The clinical manifestations of atherothrombosis in most cases are cardiac ischemia and stroke. After the acute Myocardial Infarction (MI) and stroke, venous thrombosis (VT) is most common vascular disease. Venous thrombi are produced by fibrin and red blood cells. VT and AT are associated with variety of causes. For example, VT is associated with hypercoagulatin or reduced blood flow, while AT has been linked to atherosclerosis [1].

The thrombocyte activation, which is triggered by major endothelial injury, creates main background for thrombus formation in the heart and arterial circulation. However, the high rate of blood flow prevents thrombus and clot formation. Activation and adherence of platelets are major predisposing circumstance for thrombus formation beneath the high shear stress in arteries [2]. Arterial thrombi (defined as "white") have been traditionally proposed to be composed mainly of fibrin and platelet aggregates. The Venous thrombi (defined as "red") have been proposed as mostly being enriched in fibrin and erythrocytes [3].

The significant determinant for AT is the complex interaction of lipids, inflammation and immune system, which is less, but, important for venous thrombosis as well [1]. The major mechanisms triggers intravascular thrombosis, lies at cellular and molecular levels, is known as "Virchow's triad" and includes endothelial injury, stasis or turbulent blood flow and blood hypercoagulability. Endothelial injury leading to platelet activation almost inevitably underlies thrombus formation in the heart and the arterial circulation, where the high rates of blood flow impede clot formation. Platelet adherence and activation is a necessary prerequisite for thrombus formation under high shear stress, such as exists in arteries [4].

Abnormal blood flow contributes to AT by causing endothelial injury or dysfunction. Furthermore, chaotic blood flow forms countercurrents and local pockets of stasis. Stasis is major factor in the development of venous thrombi. Under conditions of normal laminar blood flow all blood cells are found mainly in the center of the vessel lumen, separated from the endothelium by a slower moving layer of plasma, but static blood flow contributes to thrombosis. Hypercoagulability is high tendency of the blood to clot. It is an important risk factor for venous thrombosis [5].

Some peripheral blood biomarkers are useful preventive markers for arterial and venous thrombosis. The association of venous thrombosis and the ABO histo-blood group is well established. Numerous studies have demonstrated the association between the ABO blood system and venous thrombosis. It has already been frequently revealed in various populations that venous thrombosis is less frequent in the O group than in individuals with other ABO blood groups. Individuals including A, B and AB (non-O) blood group have increased of venous thrombotic disease incidence compared to the O blood group individuals.

The non-OO blood group carriers are at more significant, two-fold risk of thrombosis rather than the individuals falling into the OO blood group carriers. Accordingly, the significant serious risk of thrombosis was higher in B alleles' carriers than in those carrying A alleles [6]. Many scientific data showed the blood group's possible protective effect with venous thrombosis [7].

Under an increased risk of venous thrombosis discussed, the ABO blood group influences plasma levels of the coagulation glycoprotein named von Willebrand factor (vWF). The vWF levels are 25% lower in the O group individuals compared to other blood groups. The mechanism of how the ABO group determines plasma vWF levels has not been clearly determined. It is already known that ABO (H) carbohydrate antigenic determinants are expressed on the N-linked glycan chains of circulating plasma vWF [8].

As it already mentioned vWF is a risk factor as AT as VT. It is the molecular player in pathogenesis of thrombosis. The high level of vWF is correlated with arterial thrombosis, and especially morphological changes of vWF is a risk factor of cardiovascular disease, through promoting the platelet binding to damaged endothelium and platelet aggregation [9]. In the case of venous thrombosis, vWF is essential for flow reduction-induced thrombus formation in deep veins (Deep vein thrombosis (DVT)). The vWF-mediated platelet recruitment in DVT in the setting of blood flow disturbance plays an important pathogenic role. The gene mutations and polymorphisms, aging, hormone status, ABO blood groups, and systemic inflammation, have been involved in the modulation of both VTE predisposition and plasma levels of vWF [10].

Aging and ABO blood type influence in the bidirectional and combined way on vWF and FVIII levels. There are many studies shown that vWF significantly elevated in centenarians and type non-O individuals achieved much earlier in life the high vWF levels [11–13].

It has been investigated that the rate of venous thromboembolism is higher in black rather than white populations. Also, the non-O blood types have been associated with a notably high risk of venous thromboembolism. Since the majority of black population has an O blood and B blood type, thus may expect low cases of this disease. But it is paradoxical to find that VT prevalence has been observed to be higher in blacks than whites [14]. Due to the study to reveal suppose association between ABO and VT in blacks separately was, find out that the rate of VT was quite higher in non-O blood type individuals comparing to those with O blood groups, which also previously fixed in whites. Obtained data suggests that non-O blood type may consider as protective factor against VT in blacks too. The VT rate was higher in both, black and white non-O blood group individuals and was higher in males comparing to females and increasing with age. Further studies also confirmed above mentioned association with O group, that it may be a protective factor against VT, regardless of age and gender.

Various studies have also been reported similar to AT [15], while in different studies, this similarity was not found [6]. The study done in Switzerland reported that 2.5 times higher prevalence of B allele among MI patients than controls [16]. Underlined mechanism where may B allele is involved in pathogenesis of MI is still uncertain, since the role of ABH antigen is still unclear. In this regards various possibilities can be discussed.

The association between ABO histo-blood group, factor VIII and vWF have currently been revealed by O'Donnel and Laffan et al., [17] where BB homozygous had significantly high levels of vWF. The mechanism how B blood type affects vWF plasma levels is uncertain, but it is already defined that epitopes of N-linked oligosaccharide on the serum proteins modulate their half-lives. Hence it may be suggested that the ABH epitopes affect vWF plasma levels which cause elevation of vWF plasma levels, which is highly associated with increased risk of MI.

There are two major pathophysiological explanations tell that B antigen determines an aggregation of platelets, which may be a precursor of MI. The first mechanism is that removal of B antigen from multimeric glycoprotein by α-galactosidase decreases ristocetin cofactor activity [18] and the second is the fact which is claiming that increased platelet adhesiveness is shown in non O group individuals.

A study confirms the link between some vascular disorders and the non-O blood groups. The study has shown that there was a similar effect of ABO(H) on the level of vWF, but further research is required for obtaining more details on O(H) antigen expression levels in thrombosis. However, as A1A, A1B and BB composed of the meaningful proportion of the population attributable fraction of VT, there may be a role for more widespread adoption of ABO(H) typing in testing strategies [19]. The study of the Iranian population has not shown a significant association between ABO blood groups and coronary artery diseases. Similar results have been shown in the cases of different blood groups as well [20]. Similar results were revealed according to the study of the cardiovascular risk factors and MI in the Turkish cohort. The study showed no difference according to the cardiovascular risk factors and different blood groups within patients with MI and healthy population [21]. No association of the blood group B with MI in the sample population in Malaysia has been revealed [22].

Our research aimed to find a possible relationship between the blood groups, such as ABO, Rh, Kell, MN and AT, based on the cases of the Georgian population. The cardiovascular system diseases are also topical for Georgia, which is proved by "The European Heart Network" (EHN) (based in Brussels, EHN is an alliance of different countries together with non-governmental organizations and "Heart Foundations"). According to the research conducted by the alliance, in 2015, 29.007 deaths were reported in men and 33.509 in women ([European Cardiovascular Disease Statistics 2017 (http://www.ehnheart.org/cvd-statistics.html).

2. Research materials and methods

In our study, we aimed to determine blood group antigens in patients with AT. The study material consisted of blood samples of 100 patients with AT which were further processed according to the appropriate methodology.

The research materials were collected at the Heart Disease Department of Batumi Referral Hospital (Republic of Georgia). The research was implemented of the two years of 2019–2020. This work received financial support from the Batumi Shota Rustaveli State University (Grant project №02–12/57. 15.02.2019). Also, the control group was studied. The Control group included individuals without cardiovascular disease during the period from 2019 to 2020. They were asymptomatic, healthy persons. All participants in the control group were from the Adjara Region (Republic of Georgia). The age range of male and female individuals in both groups (patients and controls) was 40–93. All donors and patients gave their written informed consent for participation in this study, which has been duly approved by the Ethics Committee from the "UnimedAdjara" Ltd. (Adjara, Georgia).(Ethic approval number is BAH-15-1026-4/1 26.10.2015).

The express forward and reversed method with universal monoclonal antibodies was applied in order to type blood groups and reveal the blood group system antigens. Anti-A, anti-B, anti-AB monoclonal antibodies were used for ABO blood group typing (**Figure 1**). Five monoclonal antibodies, anti-D; anti-C; anti-c; anti-E and anti-e were used to identify Rh system antigens. MN blood group phenotypes have been typed by two anti-M and anti-N antibodies and Kell system antigens were determined by anti-Kell and anti-k antibodies (Bio-Rad, cypress diagnostics) (**Figure 2**). Also, for research purposes, standard O(I), (II), (III) group erythrocytes and standard O(I), A(II), B(III), AB(IV) serums were used. For identifying erythrocyte blood group antigens ID cards such as ABO/D + Reverse Grouping (Bio-Rad) (**Figure 3**) were used. The Obtained material has been studied and processed statistically.

The results were processed with the bio statistical methods. The rates between donors and patients groups were compared by Chi-square analysis of proportions.

DOI: http://dx.doi.org/10.5772/intechopen.96121

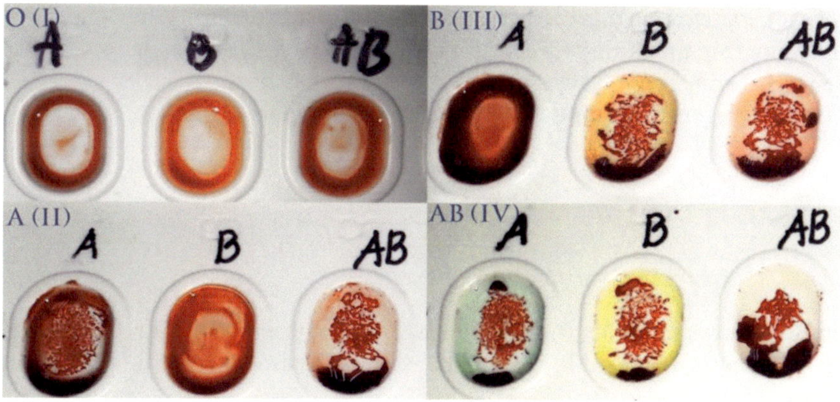

Figure 1.
ABO blood group typing by monoclonal antibodies.

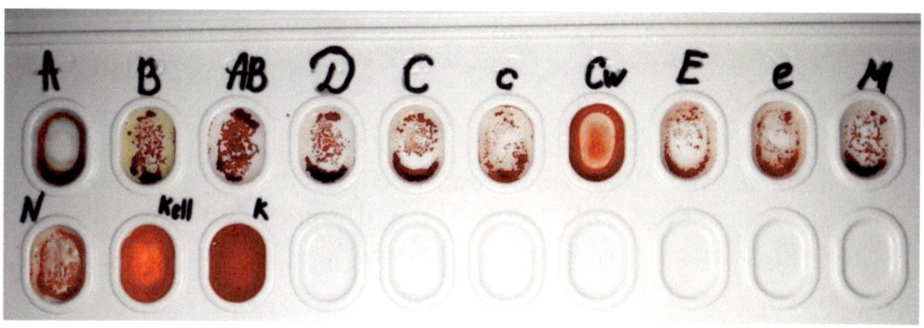

Figure 2.
ABO, Rh, MN, Kell blood group typing.

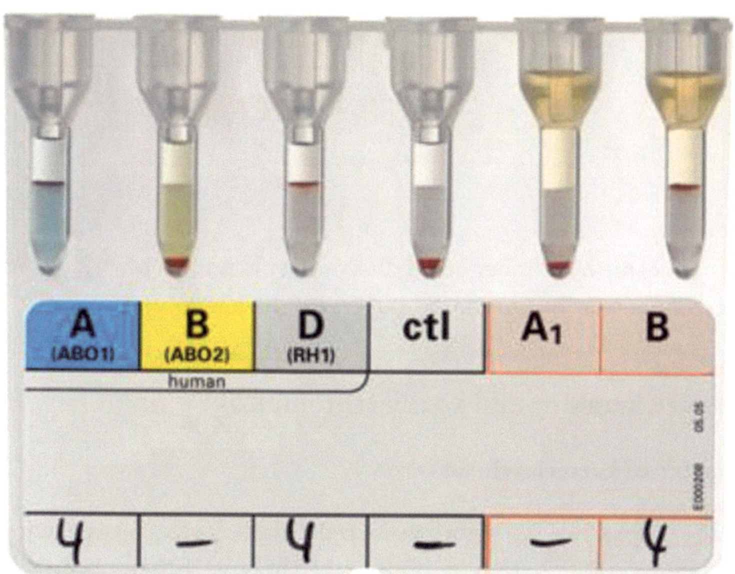

Figure 3.
ABO/D + Reverse Grouping ID card.

The level of statistical significance was set at 0.05 in all analyses. The donors and patient cohort are the same profile. All participants have the same nationality. They are Georgians and age intervals are the same for both cohort.

We used the general formula for calculating chi-square ($\chi 2$).

$$\chi_c^2 = \Sigma \frac{(O_i - E_i)^2}{E_i}$$

This test allows assessing the difference between observed (O) and expected (E) values. We used as on-variable chi-square as in some case two-variable chi-square criterion.

Gene distribution frequency of the ABO system in the studied cohorts was also analyzed. Their frequency was calculated using the formula used in the study of the three-allelic genetic system. Where O, A and B are the ration of people carrying 0 (I), A (II) and B(III) groups to the total number of research objects.

$$r = \sqrt{O};\ p = 1 - \sqrt{A+O};\ q = \mathbf{1} - \sqrt{B+O}$$

Alleles of the Kell and Rh system were also analyzed in the studied cohorts. Different frequencies of p(K), p(D) and q (k), q (d) were found in the target groups.

$$\mathbf{q} = \sqrt{\frac{n_{aa}}{N}} \quad \mathrm{p} = \mathbf{1} - \mathbf{q}$$

Where n_{aa} is a recessive homozygote according above mentioned locus (kk,dd), N is the total number of examined individuals.

The frequency of MN blood group alleles are calculated by bellow mentioned formula:

$$\mathrm{q} = \frac{n_A + \frac{1}{2} n_{AB}}{N} \quad \mathrm{p} = \frac{n_B + \frac{1}{2} n_{AB}}{N}$$

where n_A means the number of carriers of the phenotype M, n_{AB} indicates MN phenotypes, n_B means the number of carriers of the phenotype N.

3. Age, gender, smoking and arterial thrombosis

3.1 Age, gender and arterial thrombosis

As already well known, aging is considered a physiological process with the gradually ongoing accumulation of numerous structural alterations in cells, tissues, and organs. Accordingly, most alterations are associated with functional changes and have significantly increased susceptibility toward some diseases. There is an exponential increase in the risk of both arterial and venous thrombotic events with aging. This fact has been proven by several studies that revealed the increased plasma concentration of V, VII, VIII, IX factors, fibrinogen as well as vWF levels which is increasing progressively along the age [12, 23]. In addition, it has revealed that high plasma levels of fibrinogen supposedly cause the high incidence of cardiovascular events in age [24].

The activity of proteolytic enzymes (PEs) in the blood serum changes with age [25]. Physiological aging also correlates with increased plasma levels of numerous blood coagulation proteins and fibrinolysis impairment. The studies confirm that there is an association between vascular and thromboembolic diseases and aging. Thus, the state of hypercoagulability tends to increase the risk for the development of thrombosis. Accordingly, hypercoagulability considers a significant marker associated with impairment in advanced age [26]. The blood coagulation capacity increases according to age in healthy individuals because of increases in the most procoagulant factors [27]. The blood coagulation potential a gradual increase during young adulthood and an almost 2-times elevated in old age [28]. The study confirmed that aged 53–64 years is associated with significantly higher fibrinogen levels than younger aged 20 [29]. The plasma concentration of several clotting factors increase with progressing age in healthy [30]. It suggests that thrombotic diseases are found more frequently in subjects with higher plasma levels of factor VII [26].

In the human body PEs have a significant role. PEs is responsible for numerous processes in health and disease. Moreover, PEs is considered a significant factor in several physiological processes, including morphogenesis, cell differentiation, etc. [31]. According to the physiological coagulation mechanism, the calcium and platelet together contain a PEs like the trypsin, which transforms prothrombin to thrombin [32]. Paczek et al. aimed to investigate the associations between the aging process and PEs activity (including trypsin, elastase, plasmin, and active MMP-9). As they found, the active MMP-9 concentration and trypsin activity decreased according to age. Notably α 1-antitrypsin concentration and plasmin activity are elevated. This study suggests that individual PEs activity in the serum changes with age [33].

We have studied 100 patients with AT. The materials have been taken randomly. The 77% of patients were males and the rest of 23% were females. 100 volunteers were used as the controls group in the study. All volunteers were healthy. Age interval was the same for donor cohort as well. The gender ratio was not equal in patients and donors. 51% males and 49% females were studied in our controls group (**Table 1**).

As our research has shown, the gender ratios between patients and controls are different. In this case, we used a two-variable chi-square criterion. Statistically revealed a high number of chi-square criteria for both variations. This data indicates the relationships between qualitative variables. In this particular case, the value $\chi 2$ effectively rejects the null hypothesis (E = 0). The value of $\chi 2$ in the case of males is 64.32, and in the case of females equals to 39.94. These numbers are much higher than the critical value (CV) of the criterion of the degree of freedom (d.f. = 1), which is equal to 3,841. The P-Value is < .00001. The result is significant at $p < .05$ (**Table 1**).

Gender	Patients	Control	$\chi 2$, (d.f.) = 1	
Male	77 ± 4,2	51 ± 4,99	**64,32**	CV = 3,841
Female	23 ± 4,2	49 ± 4,99	**39,94**	

Table 1.
Gender ratio in donors and in the control group.

№	Age category	Patients		Control	
		Male	Female	Male	Female
1	40-49	11	0	6	3
2	50-59	10	3	6	7
3	60-69	25	4	16	20
4	70-79	23	11	17	11
5	80-89	7	4	6	6
6	90-99	1	1	0	2
n		77	23	51	49

Table 2.
Age category and gender in donors and control group.

In our study we have been allocated six age categories (**Table 2**) as well in patients as in control cohort. As it has presented on the table in young categories, age range between 40–49, there are not female patients, however from the alternative gender, 11 patients belongs to current age group. The majority of patients (56 males and 20 females) are over the 60 years old.

As our data shows, AT is much more common in males than in females. Men have a higher risk of first and recurrent venous thrombosis than women. The pathophysiology of this phenomenon is yet unknown [34]. In the United States, cardiovascular diseases are the leading cause of death; for example, in 2013 death in both sexes were equal. However, several cardiovascular diseases occur early in age-onset within men than women, like MI and coronary heart disease (CHD), where women's prevalence is ~10-year delay in onset [35–37].

3.2 Smoking and arterial thrombosis

The harmful habits such as smoking, alcohol, drugs have a negative impact on human health. In particular, they have some implications to provoke diseases of the cardiovascular system. The cigarette smoke exposure (CSE) is known to increase the risk of AT. The almost 40% of smoking-related deaths are associated with cardiovascular disease. Both active and passive CSE predispose to cardiovascular events.

It is known that cigarette smoking increases inflammation, thrombosis, and oxidation of low-density lipoproteins (LDL). Several clinical and experimental data support that cigarette smoke increases the cell's oxidative stress that can promote cardiovascular dysfunction [38]. In some studies are reported that acute CSE leads to functional changes in both fibrinogen and platelets, affecting clot dynamics, which were correlated with an alteration in fibrin structure [39].

We have analyzed the possible association between cigarette smoking and AT. Accordingly, three target groups were allocated: active smokers, non-smokers, and ex-smokers. 35% of patients were non-smokers, 39% were ex-smokers and 26% were active smokers (**Table 3**).

As it is shown from the table (**Table 3**), 35% of patients with arterial thrombosis had no exposure to tobacco use, while in the controls group nonsmoker are 1.7 times more (60%). Statistically revealed a high number of chi-square criteria, which indicates the relationships between qualitative variables. In this particular case, the value $\chi 2$ is quite effective for rejecting the null hypothesis (E = 0). The value of $\chi 2$ in the case is 55, 05. These numbers are much higher than the critical value (CV) of

Category	Smokers	Non-smoking	Ex-smokers	χ^2 (d.f. = 2)	CV
Patients	26 ± 4,38	35 ± 4,76	39 ± 4,85	55,05	5,991
Control	26 ± 4,38	60 ± 4,89	14 ± 3,46		

Table 3.
Smokers, non-smokers and ex-smoker among patients with arterial thrombosis and control.

Age	Smokers (26)		Non-smokers (35)		Ex-smokers (39)	
	Male	Female	Male	Female	Male	Female
40-49	8	0	0	0	3	0
50-59	2	0	3	2	4	1
60-69	12	0	2	4	11	0
70-79	3	0	6	10	15	1
80-89	1	0	3	4	3	0
90-99	0	0	0	1	1	0

Table 4.
Smokers, non-smokers and ex-smokers among patients with arterial thrombosis (by age and sex).

the criterion of the degree of freedom (d.f. = 2), which is equal to 5,991. The result is significant at $p < .05$.

We have also been analyzed the age and gender groups of smokers, non-smokers and ex-smokers among patients with AT (**Table 4**). As it is shown on the table there are not female patients in the smokers' (n = 26) category. All 26 active smoker patients are males. As we can see, the highest rates of AT is revealed in the category of smokers, in which only men between 60–69 years of age were observed, (12 individuals in total) and further more in the category of ex-smokers males dominates as well with the age of 70–79 (15 males and 1 female). Among whole research group 39% of patients were ex-smokers.

It is a wonder why the number of ex-smokers exceeds to the active smokers in our research? We do not have sufficient information when and with what a reason ex-smoker patient group gave up smoking. The high prevalence of ex-smokers in patients may be due to the fact that they had to quit smoking due to deterioration their health condition. On the other side the ex-smokers have a greater experience of smoking (at least 10 years), they have accumulated more toxic substances in the body, atherosclerotic plaques, have abnormal fibrin structure and etc., which may also be one of the considerable reasons for high incidence [40, 41].

4. Blood groups and arterial thrombosis

4.1 ABO blood group and arterial thrombosis

Many studies have found that blood group phenotypes play as an important genetic risk factor in many diseases. For example, in hemostasis, the ABO blood group has a great influence since it is a crucial determinant of the vWF and therefore, to coagulation factor VIII (FVIII). Besides, an individual with O blood group has around 25% lower vWF and FVIII circulating levels comparing to other blood group individuals. Some experimental studies show that several inflammatory

Category	O (I)	A (II)	B (III)	AB (IV)	ABO system Genes frequency			χ^2 (d.f. = 3)	CV
					r	p	q		
Patients	47 ± 4,99	38 ± 4,83	13 ± 3,36	2 ± 1,4	0,69	0,08	0,23	0,09	7,815
Control	48 ± 4,99	3 8 ± 4,83	12 ± 3,24	2 ± 1,4					

Table 5.
ABO blood group phenotypes distribution among patients with AT and control.

cytokines and cholesterol levels may be the most likely mechanisms to explain an association between ABO blood group and cardiovascular diseases [42–45].

Numerous studies have established an association between venous thrombosis and the non-0 group. The mechanism of an association with the ABO blood group and cardiovascular disease has not been fully explored. However, there is a hypothesis that ABO glycosyltransferases modify glycoproteins and glycolipids on platelets' surface effects on the platelet function. The antigens of the ABO system are not presented only on the erythrocyte surface; There are also founded on the cells of epithelium platelet and vascular endothelium, as well [46]. Thus, they are extending potential pathophysiology into other areas of cardiovascular disease and postoperative outcomes.

According to the authors, it is suggested that venous and arterial thrombotic disorders have different pathophysiological backgrounds; for example, arterial thrombosis has extended by platelet activation, while venous thrombosis is based on the clotting system activation [3].

Our study aimed to find an association between blood groups and arterial thrombosis. 100 patients with arterial thrombosis and the same number (100) healthy control group's were investigated. In the control group, the distribution of ABO blood group phenotypes was as follows: group O – 48%; group A – 38%; group B –12%; and group AB – 2%. A similar distribution was revealed in the patients as well (**Table 5**). As the obtained data confirms, the correlation was not found between the ABO blood group and AT. The prevalence of ABO blood group phenotypes (O, A, B, AB) is the same in the patients and the controls. ABO system gene frequency is the same both cohort. The r allele frequency is 0,68, p – 0,08 and q – 0,23.

In this case, we have a Null hypothesis (E = 0). This means that there is no association between the two variables. The value of $\chi 2$ criteria is equal to 0,09, which is much less than the critical (CV) value of the degree of freedom (d.f. = 3), equal to 7,815. The P-Value is.029291. The result is significant at $p < .05$. Our result is the same as the study, including Iranian patients with coronary artery disease. The association with particular ABO blood groups was not found based on the comparison analysis between coronary artery disease general population [20]. The same data is in the study of risk factors and myocardial infarction (MI) in a Turkish cohort. The Turkish population study shows that ABO blood group antigens' frequency was the same in the patients and the control group [21]. There is no association of blood group B with MI in the sample population in Malaysia [22].

4.2 Rh factor and arterial thrombosis

Our aim was also to find a possible correlation between Rh blood group antigens and AT. Next to the ABO system, the Rh blood group system is the second significant human blood group system [47], which comprises 49 blood group antigens, among them the five D, C, c, E, and e antigens are the clinically most important

DOI: http://dx.doi.org/10.5772/intechopen.96121

Category	Rh^+	Rh^-	Rh system gene frequency		χ^2 (d.f. = 1)	CV
			q	p		
Patients	78	22	0,46	0,54	0,05	3,841
Control	79	21				

Table 6.
Distribution of Rh positive and Rh-negative blood group in patients and control.

antigens. Rh blood group is associated with the hemolytic disease of the newborn. The hemolytic condition occurs when there is an incompatibility between the mother and the fetus's Rh blood types.

The majority of the available studies involved ABO blood group and risk of cardiovascular diseases, while there is a lack of data on the Rh factor's effect on MI's risk and prognosis. Serum cholesterol varied significantly in the Rh system, but not in the AB0 system [48].

The Rh family proteins include none erythroid Rh homologs found on the epithelial tissues of different organs, such as kidney, liver, brain, and skin [49]. Some data points to the coincidence of certain cardiovascular risk factors with the Rh blood group [50]. It has been described that positive Rh is an independent risk factor for low levels of high-density lipoprotein (HDL) cholesterol, which is a well-known risk factor for cardiovascular disorders. Except for that, several clinical studies reported that the Rh genotype is correlated with systolic blood pressure [51].

Rh factor was studied in both patients and control cohort. 78 studied patients were Rh-positive and the remaining 22 were Rh-negative. The prevalence of the Rh phenotypes was similar in the control group (79 – Rh-positive and 21 - Rh-negative) (**Table 6**). Rh system gene frequency is the same for both cohorts. The frequency of q allele is 0,46, p allele – 0,54.

In our study, the prevalence of the Rh blood group phenotypes (Rh-negative and Rh-positive) is the same in the patients and the controls. Based on our study, we could not found any correlation between arterial thrombosis and Rh blood group because the Rh phenotypes in the controls and patients groups are equally distributed. We are not able to reject the Null hypothesis (E = 0). There is no reliable difference between the two categories. The value of $\chi 2$ criteria, in this case, is equal to 0.05, which is much less than the Critical Value (CV) of the degree of freedom (d.f. = 1), which is equal to 3,841. The P-Value is.025347. The result is significant at $p < .05$ (**Table 6**).

4.3 MN blood group and arterial thrombosis

The MNS antigen system is a human blood group system based upon two genes (glycophorin A and glycophorin B) on the 4th chromosome 4q28.2-q13.1. There are currently 46 antigens in the system [52], but the five most important are called M, N, S, s, and U. We have studied 2 antigens from this system. Based on this, we have allocated 3 phenotypes (M group, N group, and MN group) in the control cohort and the patients. There is little information about MN blood group correlation with some kind of infectious disease [53, 54].

No information could have been obtained through the scientific literature about the correlation between MN blood groups and AT. But there are some information about association of MN blood group and hypertension. Males with the MN phenotype had significantly higher unadjusted systolic and diastolic blood pressures than those who were homozygous MM or NN [55]. The increased of

Category	M+N+	M+N-	M-N+	MN allele frequency		χ^2, (d.f. = 2)	CV
				q	P		
Patients	43	48	9	0,695	0,305	8,95	5,991
Control	30	55	15	0,7	0,3		

Table 7.
MN group system phenotype frequency in patients and control group.

erythrocyte sodium-lithium counter-transport activity associated with hypertension. Erythrocyte sodium-lithium counter-transport genetically is heritable, but it is not monogenic inheritance [56]. It is complex genetically characteristics. Some authors studied the association of blood pressure, sodium-lithium counter-transport and two genetic markers – MN Blood group system and haptoglobin in Michigan population example. They find genetic linkage between the MN locus and red blood cell sodium-lithium counter-transport activity. The authors suggested that the relation between MN phenotype and systolic blood pressure is different in men and women. The men with the MN phenotype and having high systolic blood pressure had significantly elevated red blood cell sodium-lithium counter-transport activity [57].

Delanghe et al., study suggested that detection age of hypertension was lower for patients with MN phenotype. The increasing age of detection the distribution of MN phenotype gradually decreases. The hypertensive with MM blood group had a lower prevalence of cerebrovascular accidents [58].

Sobha R. et al., found that the distribution of MN blood groups differs significantly in asthma patients and controls (chi-square - 15.1160, d.f. 2, Pp > .2) suggesting that blood group MM is resistant to asthma. The comparison of NN blood type reveals non-significant differences between asthma patients and controls [59].

In our studies the prevalence of MN blood group phenotypes was not equal in controls and patients groups (**Table 7**). The frequency of M+ N+ (MN) phenotype was relatively high in the patient groups compared to the controls. 43/100 patients and 30/100 donors carry the M+ N+ (MN) phenotype. In contrast, the major part of the control group (n = 55) had M+ N- (MM) phenotype, which is 1,14 times higher than in the patients' group (n = 48). The study also has revealed the differences according to the prevalence of the M-N+ (NN) phenotype control and patients' groups. The nine patients had the M-N+ phenotype characteristics, which is 1,8 times less than in the donors (n = 15). The study of MN system alleles frequency do not show any significant differences between to cohorts (**Table 7**).

The $\chi 2$ criteria is 8,95, which is only 3 units higher then CV (d.f. = 3). In this case, the chi-square criterion indicates the existence of relationship between the two qualitative variables and the rejection of the null hypothesis (E = 0). The P-Value is < .00001. The result is significant at p < .05 (**Table 7**).

4.4 Kell blood group system and arterial thrombosis

The Kell blood group system is a complex and contains many highly immunogenic antigens. The locus of Kell blood group system is highly polymorphic and gives rise to many Kell antigens in human being. The Kell system gene locus is found on chromosome 7, at 7q33 and contains 19 exons. This locus encoding the 25 antigens and majority of Kell blood group system antigens are glycoproteins. Some of them are highly immunogenic. The Kell antigen is expresses only by erythroid progenitor cells and mature erythroid cells. There are two codominant allelic genes that produce two main antigens: Kell (K) and Cellano (k). These antigens differ to

Category	k	Kell	Kell system alleles frequency		χ^2, (d.f. = 1)	CV
			q	p		
Patients	79	21	0,88	0,12	51,73	3,841
Control	87	13	0.93	0,07		

Table 8.
Kell and k phenotype frequency in patients and control group.

each other by a single amino acid. The prevalence of these antigens is not same. The k antigen is more common in most world population. K–K+ phenotype is found in 98% of Blacks and 91% of Caucasians. These antigens are the third most potent antigens after ABO and Rh blood groups, triggering an immune reaction [60].

Kell blood group system is important not only transfusion but also clinically is associated with the Hemolytic Disease of the Newborn (HDN). HDN caused by anti-Kell1 antibody is the second most common after Rh disease. Anti-Kell is becoming relatively more important as the prevention of Rh disease is also becoming more effective.

Unlikely to Rhesus and ABO immunosensibilization, HDN attributable to Kell sensitization is causes by anti-K antibody suppressing the fetal red blood cell (RBCs) production. The main reason of this is well known. The Kell antigens generally expressed on the surface of RBCs precursors, and anti-K antibodies initiated immune destruction of K positive elytroid precursor cells by macrophages in the fetal liver. The RBCs progenitor cells do not contain hemoglobin. There is releasing less amount of bilirubin and newborn very rarely presents a jaundice phenotype of anemia [61].

We could not have obtained any information about the correlation between the Kell blood group antigens and cardiovascular disease. The study according to the Kell system 2 phenotypical groups (K+ and K- phenotypes), showed that the frequency of K+ phenotype (KK or Kk) is a high in the patients compared to the controls. 87 of the studied control have the K-negative (kk) phenotype, and as for the patients with arterial thrombosis, the prevalence of the above-mentioned phenotype is 79. The Kell antigen prevalence was relatively high in the studied patients (n = 21) and 13 healthy control were carriers of Kell antigen. K+ phenotype characteristics are highly prevalent in patients than in donors. Kell system alleles frequency are not same between to cohort (**Table 8**). In this case, quite a high number of $\chi 2$ criteria was observed, which indicates the relationship between qualitative variables. $\chi 2$ value is equal to 51,73. Which is much more than CV (3,841) of df (d.f. = 1). The P-Value is < .00001. The result is significant at < .05 (**Table 8**).

5. Conclusion

Thus, our study has shown maleness as a higher risk factor for AT. The study also revealed that smokers have a more predicted chance for AT compared to non-smokers. K+ phenotype and M+ N+ characteristics are of the high prevalence in the patients than in the donors. The correlations of the Kell and MN blood groups with AT have practical significance. There is no correlation between the ABO and Rh blood groups with AT based on our research; however, we think that finding the possible association needs to be identified with multicenter, prospective, and large-scale studies. Based on our research, it is possible to single out high-risk groups of disease and implement preventive arrangements on the individuals of these groups.

Acknowledgements

This work received financial support from the Batumi ShotaRustaveli State University. Targeted grant project №02-12/57. 15.02.2019.

Conflict of interest

The authors declare no conflict of interest.

Author details

Marina Koridze[1], Marina Nagervadze[1*], Maria Sarkhaiani[1], Leila Akhvlediani[2], Rusudan Khukhunaishvili[1], Ketevan Dolidze[1], Sophiko Tskvitinidze[1], Shorena Gabaidze[1], Irina Nakashidze[1], Sopio Garakanidze[1] and Giorgi Nikolaishvili[1]

1 Batumi Shota Rustaveli State University, Batumi, Georgia

2 BAU International University Batumi, Batumi, Georgia

*Address all correspondence to: nagervadze.marina@bsu.edu.ge

References

[1] Previtali E, Paolo B. Risk factors for venous and arterial thrombosis. Blood Transfus. 2011;120-38.

[2] Endothelial cell control of thrombosis [Internet]. [cited 2021 Jan 7]. Available from: https://www.ncbi.nlm.nih.gov/pmc/articles/PMC4617895/

[3] Prandoni P. Venous and arterial thrombosis: Two aspects of the same disease? Clin Epidemiol. 2009 Jan;1.

[4] Vascular bed-specific thrombosis - AIRD - 2007 - Journal of Thrombosis and Haemostasis - Wiley Online Library [Internet]. [cited 2021 Jan 15]. Available from: https://onlinelibrary.wiley.com/doi/full/10.1111/j.1538-7836.2007.02515.x

[5] Chiu J-J, Chien S. Effects of Disturbed Flow on Vascular Endothelium: Pathophysiological Basis and Clinical Perspectives. Physiol Rev. 2011 Jan;91(1):327-87.

[6] Jukić I, Bingulac-Popović J, Đogić V, Babić I, Culej J, Tomičić M, et al. ABO Blood Groups and Genetic Risk Factors for Thrombosis in Croatian Population. Croat Med J. 2009 Dec;50(6):550-8.

[7] de Paula Sabino A, Ribeiro DD, Domingheti CP, Rios DRA, Dusse LMS, das Graças Carvalho M, et al. ABO blood group polymorphisms and risk for ischemic stroke and peripheral arterial disease. Mol Biol Rep. 2014 Mar;41(3):1771-7.

[8] Jenkins PV, O?Donnell JS. ABO blood group determines plasma von Willebrand factor levels: a biologic function after all? Transfusion (Paris). 2006 Oct;46(10):1836-44.

[9] Variation in the von Willebrand factor gene is associated with von Willebrand factor levels and with the risk for cardiovascular disease | Blood | American Society of Hematology [Internet]. [cited 2021 Jan 15]. Available from: https://ashpublications.org/blood/article/117/4/1393/28525/Variation-in-the-von-Willebrand-factor-gene-is

[10] Brill A, Fuchs TA, Chauhan AK, Yang JJ, De Meyer SF, Köllnberger M, et al. von Willebrand factor–mediated platelet adhesion is critical for deep vein thrombosis in mouse models. Blood. 2011 Jan 27;117(4):1400-7.

[11] Albánez S, Ogiwara K, Michels A, Hopman W, Grabell J, James P, et al. Aging and ABO blood type influence von Willebrand factor and factor VIII levels through interrelated mechanisms. J Thromb Haemost. 2016 May;14(5):953-63.

[12] Von Willebrand factor in Italian centenarians - PubMed [Internet]. [cited 2021 Jan 3]. Available from: https://pubmed.ncbi.nlm.nih.gov/12551825/

[13] Zhou S. Is ABO blood group truly a risk factor for thrombosis and adverse outcomes? World J Cardiol. 2014;6(9):985.

[14] Fang C, Cohen HW, Billett HH. Race, ABO blood group, and venous thromboembolism risk: not black and white: RACE, ABO BLOOD GROUP, AND VTE RISK. Transfusion (Paris). 2013 Jan;53(1):187-92.

[15] Capuzzo E, Bonfanti C, Frattini F, Montorsi P, Turdo R, Previdi MG, et al. The relationship between ABO blood group and cardiovascular disease: results from the Cardiorisk program. Ann Transl Med. 2016 May;4(10):189-189.

[16] Nydegger UE, Wuillemin WA, Julmy F, Meyer BJ, Carrel TP. Association of ABO histo-blood group B allele

with myocardial infarction: Histo-blood group B allele associated with myocardial infarction. Eur J Immunogenet. 2003 Jun;30(3):201-6.

[17] The relationship between ABO histo-blood group, factor VIII and von Willebrand factor - O'Donnell - 2001 - Transfusion Medicine - Wiley Online Library [Internet]. [cited 2021 Jan 15]. Available from: https://onlinelibrary.wiley.com/doi/abs/10.1046/j.1365-3148.2001.00315.x

[18] Role of A and B blood group antigens in the expression of adhesive activity of von Willebrand factor - Sarode - 2000 - British Journal of Haematology - Wiley Online Library [Internet]. [cited 2021 Jan 15]. Available from: https://onlinelibrary.wiley.com/doi/full/10.1046/j.1365-2141.2000.02113.x

[19] Wu O, Bayoumi N, Vickers MA, Clark P. ABO(H) blood groups and vascular disease: a systematic review and meta-analysis: ABO groups and thrombosis. J Thromb Haemost. 2007 Oct 25;6(1):62-9.

[20] Amirzadegan A, Salarifar M, Sadeghian S, Davoodi G, Darabian C, Goodarzynejad H. Correlation between ABO blood groups, major risk factors, and coronary artery disease. Int J Cardiol. 2006 Jun;110(2):256-8.

[21] Sari I, Ozer O, Davutoglu V, Gorgulu S, Eren M, Aksoy M. ABO blood group distribution and major cardiovascular risk factors in patients with acute myocardial infarction: Blood Coagul Fibrinolysis. 2008 Apr;19(3):231-4.

[22] Association of ABO blood group B with myocardial infarction - PubMed [Internet]. [cited 2021 Jan 3]. Available from: https://pubmed.ncbi.nlm.nih.gov/19651016/

[23] Franchini M. Hemostasis and aging. Crit Rev Oncol Hematol. 2006 Nov;60(2):144-51.

[24] Ofosu FA, Craven S, Dewar L, Anvari N, Andrew M, Blajchman MA. Age-related changes in factor VII proteolysis *in vivo*. Br J Haematol. 1996 Aug;94(2):407-12.

[25] Ritthaler U, Deng Y, Zhang Y, Greten J, Abel M, Sido B, et al. Expression of Receptors for Advanced Glycation End Products in Peripheral Occlusive Vascular Disease. Am J Pathol [Internet]. 1995 Mar [cited 2021 Jan 14];146(3):688-94. Available from: https://www.ncbi.nlm.nih.gov/pmc/articles/PMC1869189/

[26] Mari D, Ogliari G, Castaldi D, Vitale G, Bollini EM, Lio D. Hemostasis and ageing. Immun Ageing [Internet]. 2008 Dec [cited 2021 Jan 14];5(1):12. Available from: https://immunityageing.biomedcentral.com/articles/10.1186/1742-4933-5-12

[27] Kurachi K, Kurachi S. Genetic Mechanisms of Age Regulation of Blood Coagulation: Factor IX Model. Arterioscler Thromb Vasc Biol [Internet]. 2000 Apr [cited 2021 Jan 14];20(4):902-6. Available from: https://www.ahajournals.org/doi/10.1161/01.ATV.20.4.902

[28] Favaloro E, Franchini M, Lippi G. Aging Hemostasis: Changes to Laboratory Markers of Hemostasis As We Age—A Narrative Review. Semin Thromb Hemost [Internet]. 2014 Aug 6 [cited 2021 Jan 14];40(06):621-33. Available from: http://www.thieme-connect.de/DOI/DOI?10.1055/s-0034-1384631

[29] Meade TW, North WRS, Chakrabarti R, Haines AP, Stirlig Y. POPULATION-BASEDDISTRIBUTIONS OF HAEMOSTATIC VARIABLES. Br Med Bull [Internet]. 1977 Sep [cited 2021 Jan 14];33(3):283-8. Available

from: https://academic.oup.com/bmb/article/265033/POPULATION-BASED

[30] Mari D, Coppola R, Provenzano R. Hemostasis factors and aging. Exp Gerontol [Internet]. 2008 Feb [cited 2021 Jan 14];43(2):66-73. Available from: https://linkinghub.elsevier.com/retrieve/pii/S0531556507001404

[31] Cyganek A, Wyczalkowska-Tomasik A, Jarmuzek P, Grzechocinska B, Jabiry-Zieniewicz Z, Paczek L, et al. Activity of Proteolytic Enzymes and Level of Cystatin C in the Peripartum Period. BioMed Res Int [Internet]. 2016 [cited 2021 Jan 14];2016:1-5. Available from: https://www.hindawi.com/journals/bmri/2016/7065821/

[32] Eagle H, Harris T. Coagulation of Blood by Proteolytic Enzymes (Trypsin, Papain). Exp Biol Med [Internet]. 1936 Oct 1 [cited 2021 Jan 14];35(1):157-8. Available from: http://ebm.sagepub.com/lookup/doi/10.3181/00379727-35-8891P

[33] Paczek L, Michalska W, Bartlomiejczyk I. Trypsin, elastase, plasmin and MMP-9 activity in the serum during the human ageing process. Age Ageing [Internet]. 2008 Mar 10 [cited 2021 Jan 14];37(3):318-23. Available from: https://academic.oup.com/ageing/article-lookup/doi/10.1093/ageing/afn039

[34] Roach REJ, Cannegieter SC, Lijfering WM. Differential risks in men and women for first and recurrent venous thrombosis: the role of genes and environment. J Thromb Haemost. 2014 Oct;12(10):1593-600.

[35] Maas AHEM, Appelman YEA. Gender differences in coronary heart disease. Neth Heart J. 2010 Nov;18(12):598-603.

[36] Kyrle PA, Minar E, Bialonczyk C, Hirschl M, Weltermann A, Eichinger S. The Risk of Recurrent Venous Thromboembolism in Men and Women. N Engl J Med. 2004 Jun 17;350(25):2558-63.

[37] Mozaffarian D, Benjamin EJ, Go AS, Arnett DK, Blaha MJ, Cushman M, et al. Executive Summary: Heart Disease and Stroke Statistics—2016 Update: A Report From the American Heart Association. Circulation. 2016 Jan 26;133(4):447-54.

[38] Armani C, Landini Jr. L, Leone A. Molecular and Biochemical Changes of the Cardiovascular System due to Smoking Exposure. Curr Pharm Des. 2009 Apr 1;15(10):1038-53.

[39] Barua RS, Sy F, Srikanth S, Huang G, Javed U, Buhari C, et al. Effects of Cigarette Smoke Exposure on Clot Dynamics and Fibrin Structure: An Ex Vivo Investigation. Arterioscler Thromb Vasc Biol. 2010 Jan;30(1):75-9.

[40] Violi F, Pastori D, Pignatelli P, Carnevale R. Nutrition, Thrombosis, and Cardiovascular Disease. Circ Res. 2020 May 8;126(10):1415-42.

[41] Loffredo L, Perri L, Nocella C, Violi F. Antioxidant and antiplatelet activity by polyphenol-rich nutrients: focus on extra virgin olive oil and cocoa: Antioxidant and antiplatelet activity of extra virgin olive oil and cocoa. Br J Clin Pharmacol. 2017 Jan;83(1):96-102.

[42] Paré G, Chasman DI, Kellogg M, Zee RYL, Rifai N, Badola S, et al. Novel Association of ABO Histo-Blood Group Antigen with Soluble ICAM-1: Results of a Genome-Wide Association Study of 6,578 Women. Gibson G, editor. PLoS Genet. 2008 Jul 4;4(7):e1000118.

[43] Karakas M, Baumert J, Kleber ME, Thorand B, Dallmeier D, Silbernagel G, et al. A Variant In the Abo Gene Explains the Variation in Soluble E-Selectin Levels—Results from Dense Genotyping in Two Independent

Populations. Arking DE, editor. PLoS ONE. 2012 Dec 28;7(12):e51441.

[44] Paterson AD, Lopes-Virella MF, Waggott D, Boright AP, Hosseini SM, Carter RE, et al. Genome-Wide Association Identifies the ABO Blood Group as a Major Locus Associated With Serum Levels of Soluble E-Selectin. Arterioscler Thromb Vasc Biol. 2009 Nov;29(11):1958-67.

[45] Chen Y, Chen C, Ke X, Xiong L, Shi Y, Li J, et al. Analysis of Circulating Cholesterol Levels as a Mediator of an Association Between ABO Blood Group and Coronary Heart Disease. Circ Cardiovasc Genet. 2014 Feb;7(1):43-8.

[46] Franchini M, Rossi C, Mengoli C, Frattini F, Crestani S, Giacomini I, et al. ABO blood group and risk of coronary artery disease. J Thromb Thrombolysis. 2013 Oct;36(3):286-7.

[47] Avent ND, Reid ME. The Rh blood group system: a review. Blood. 2000 Jan 15;95(2):375-87.

[48] Korsan-Bengtsen K, Wilhelmsen L, Nilsson L-Å, Tibblin G. Blood coagulation and fibrinolysis in relation to ABO, Rh, MN and Duffy blood groups in a random population sample of men aged 54 years. Thromb Res. 1972 Dec;1(6):549-58.

[49] The Structure and Function of the Rh antigen Complex [Internet]. [cited 2021 Jan 3]. Available from: https://www.ncbi.nlm.nih.gov/pmc/articles/PMC1831834/

[50] Kanbay M, Yildirir A, Ulus T, Bilgi M, Kucuk A, Muderrisoglu H. Rhesus Positivity and Low High-Density Lipoprotein Cholesterol: A New Link? Asian Cardiovasc Thorac Ann. 2006 Apr;14(2):119-22.

[51] Robinson MT, Wilson TW, Nicholson GA, Grell GAC, Etienne C, Grim CM, et al. AGT and RH blood group polymorphisms affect blood pressure and lipids in Afro-Caribbeans. J Hum Hypertens. 2004 May;18(5):351-63.

[52] Daniels G, Flegel WA, Fletcher A, Garratty G, Levene C, Lomas-Francis C, et al. International Society of Blood Transfusion Committee on Terminology for Red Cell Surface Antigens: Cape Town report. Vox Sang. 2007 Apr;92(3):250-3.

[53] Acosta O, Solano L, Escobar J, Fernandez M, Solano C, Fujita R. Frequencies of Blood Group Systems MNS, Diego, and Duffy and Clinical Phases of Carrion's Disease in Amazonas, Peru. Interdiscip Perspect Infect Dis. 2014;2014:1-8.

[54] Blood Groups in Infection and Host Susceptibility [Internet]. [cited 2021 Jan 3]. Available from: https://www.ncbi.nlm.nih.gov/pmc/articles/PMC4475644/

[55] Blood pressure and blood group markers: association with the MN locus - PubMed [Internet]. [cited 2021 Jan 16]. Available from: https://pubmed.ncbi.nlm.nih.gov/6597819/

[56] Weder AB, Delgado MC, Zhu X, Gleiberman L, Kan D, Chakravarti A. Erythrocyte Sodium-Lithium Countertransport and Blood Pressure: A Genome-Wide Linkage Study. Hypertension. 2003 Mar;41(3):842-6.

[57] Weder AB, Schork NJ, Julius S. Linkage of MN locus and erythrocyte lithium-sodium countertransport in Tecumseh, Michigan. Hypertension. 1991 Jun;17(6_pt_2):977-81.

[58] Delanghe J, Duprez D, Buyzere MD, Robbrecht D, Bergez B, Leroux-Roels G, et al. MN blood group, a genetic marker for essential arterial hypertension in young adults. Eur Heart J. 1995 Sep;16(9):1269-76.

[59] ABO, Rh & MN Blood Groups in Relation to Asthma on JSTOR [Internet]. [cited 2021 Jan 16]. Available from: https://www.jstor.org/stable/41919619?seq=1

[60] The Blood Group Antigen FactsBook - 3rd Edition [Internet]. [cited 2021 Jan 3]. Available from: https://www.elsevier.com/books/the-blood-group-antigen-factsbook/reid/978-0-12-415849-8

[61] Vaughan JI, Manning M, Warwick RM, Letsky EA, Murray NA, Roberts IAG. Inhibition of Erythroid Progenitor Cells by Anti-Kell Antibodies in Fetal Alloimmune Anemia. N Engl J Med. 1998 Mar 19;338(12):798-803.